Practical   Part One
Pre-Algebra

Nathan Y Ko & Kwang S. Ko, Ph.D.

1   *A SELF-STUDY GUIDE*
2   *EXERCISES*
3   *SELF-TESTS*
4   *QUICK EXERCISES*
5   *FULL ANSWER KEY*

Part One

CONTENTS

Part Two*

* The following is a separate book [Practical Pre-Algebra (part two)].

CONTENTS

CHAPTER 1
Basic Concepts of Algebra

You will learn basic concepts of algebra and number sense based on the number system, algebraic expressions, and order of operations.

1. Number System

1–1. Number system

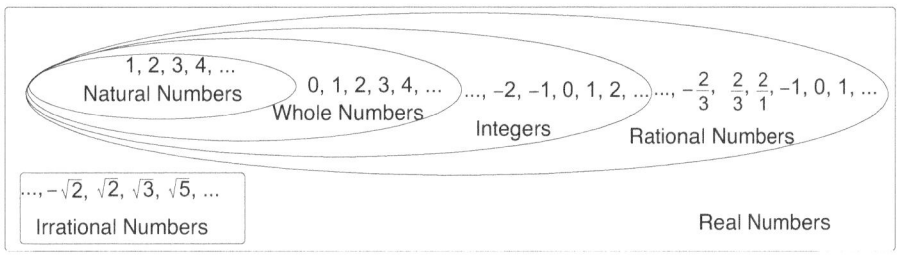

1) Natural numbers are whole numbers from 1 upwards: 1, 2, 3, 4, ...
2) Whole numbers are simply the numbers from 0 upwards: 0, 1, 2, 3, ...
3) Integers are similar to whole numbers but they also include negative numbers: . . . , −2, −1, 0, 1, 2, . . .
4) Rational numbers are real numbers that can be made by dividing two integers and includes integers, terminating decimals, and repeating decimals.
5) Irrational numbers are real numbers that cannot be written as simple fractions and have nonterminating decimals.
6) Real numbers are all the numbers on a number line.

1–2. Name the number set that each value belongs to.

a) $1.2424\cdots$ b) -7 c) $\sqrt{81}$ d) $\{3, -3, \frac{1}{3}, \sqrt{3}\}$

SOLUTION

a) $1.2424\cdots$ is a rational number because it has a repeating decimal.
b) -7 is an integer and rational number.
c) $\sqrt{81}$ is a perfect square root that equals to 9. It is a natural number, a whole number, an integer and a rational number.
d) $\{3, -3, \frac{1}{3}, \sqrt{3}\}$ are a set of real numbers.

Quick Exercises 1   Name the number set(s) that each value belongs to.

1) $\sqrt{100}$

2) $\sqrt{3}$

3) 0

4) $-\dfrac{1}{3}$

1–3. What is the phrase equivalent to the algebraic expression below?

$$2x^3$$

SOLUTION

2, x, and 3 are in the expression. In the expression, x^3 is called "x to the third power", meaning that "$x \times x \times x$".

$$\underset{\text{base}}{\overset{\text{exponent}}{2x^3}} = 2 \text{ times } x \text{ to the } \underline{\text{third power}}$$
$$\text{(cubed)}$$

So $2x^3$ is equivalent to "$2 \times x \times x \times x$".
However, you should know the following are not true. i) $2x^3 \neq 2(3^x)$, ii) $2x^3 \neq 3(2^x)$, iii) $2x^3 \neq 2(x + x + x)$, and iv) $2(x \times x \times x) \neq 2(3x)$.

1–4. Find the value of the expression.

$$4^2 + 2^4$$

SOLUTION

In the expression, 4^2 is called "4 squared or 4 to the second power", meaning that "4×4" and 2^4 is called "2 to the fourth power", meaning that "$2 \times 2 \times 2 \times 2$".
$4^2 = 4 \times 4 = 16$, $2^4 = 2 \times 2 \times 2 \times 2 = 16$
So, the value of $4^2 + 2^4$ is 32.

Quick Exercises 2   Write out each phrase or algebraic expression

1) y^3x^4

2) 2^3x^2

3) x to the seventh power

4) 4 times 2 to the fourth power

2. Order of Operations

1–5. Find the value of the expression.

$$5 - 3^2 \times 2 + (7 - 5) \div 2$$

SOLUTION

When solving an expression with more than one operation, there is a standard order.

Order of the Operations System
i) First, perform any calculations inside the parentheses.
ii) Second, solve the exponents.
iii) Third, perform all multiplications and division problems, working from left to right.
iv) Finally, perform all addition and subtraction problems, working from left to right.

$5 - 3^2 \times 2 + (7 - 5) \div 2$	Original expression.
$5 - 3^2 \times 2 + 2 \div 2$	Calculate the problems in the parentheses.
$5 - 9 \times 2 + 2 \div 2$	Solve the exponents.
$5 - 18 + 1$	Multiply and divide from left to right.
-12	Subtract and add from left to right.

So, the value of the expression is -12.

Quick Exercises 3   Find the value of each expression.

1) $(1 + 2) \cdot 2^2 \div 2 - 1$

2) $30 \div (2 \cdot 3) - 8 \div 2^2$

3) $(37 - 1) \div [5 + (8 - 4)] \times 4$

4) $2^5 \div [(8 - 4) \times 2]$

1–6. If $x = 2$, find the value of $4^2(7 - 6) + 8 \div x$.

SOLUTION

When you look at the expression, there are more than one operation and a variable. You should know what x represents in order to solve the problem. In this case, $x = 2$. So you may insert the value of x first, then solve the expression using the order of operations.

$4^2(7 - 6) + 8 \div x$	Original expression.
$4^2(7 - 6) + 8 \div 2$	Replace 2 with x.
$4^2 + 8 \div 2$	Compute in parentheses. $(7 - 6) = 1$
$16 + 8 \div 2$	Compute the exponential.
$16 + 4$	Divide 8 by 2.
20	Add 16 and 4.

So, the value of the expression is 20.

Quick Exercises 4   Find the value of each expression.

1) If $w = 4$, find the value of $(8 \times w) - 2.9$.   2) If $k = 46$, find the value of $2(k \div 2) + 1$.

3) If $\dfrac{6}{x} - 5 = 2$, find the value of $2^2 + (6 \div x)$.   4) If $c \div 2 = 6$, find the value of $\dfrac{(c \div 2) + 2}{2}$.

1–7. Find the value of $\dfrac{2(3^2 - 2^3)}{2 \cdot 3^2}$.

SOLUTION

$$\underbrace{[2(3^2 - 2^3)]}_{\text{dividend}} \div \underbrace{(2 \cdot 3^2)}_{\text{divisor}} \;=\; \overset{\text{numerator}}{\dfrac{2(3^2 - 2^3)}{2 \cdot 3^2}}_{\text{denominator}}$$

Follow the order of operations.

$\dfrac{\cancel{2}(3^2 - 2^3)}{\cancel{2} \cdot 3^2}$ Original expression

$\dfrac{(3^2 - 2^3)}{3^2}$ Cancel out 2 and 2.

$\dfrac{(9 - 8)}{9}$ Solve the problem within the parentheses.

$\dfrac{1}{9}$ Subtract.

So, the value of the expression is $\dfrac{1}{9}$.

Quick Exercises 5   Find the value of each expression.

1) $\dfrac{4^3 - 2^2}{3^2 + 1^2}$ 2) $\dfrac{2(2^3 + 4^2)}{(2)(2^3)}$

3) $4(2^2 + 1) \div 2(4 \div 2)$ 4) $2(6 \div 2) \div [(4 \div 2)^2 + 2]$

Exercises 1 Write out each phrase or algebraic expression. Simplify if necessary.

1) 2^4

2) x^5

3) $5 \cdot 2^3$

4) $3 \cdot a^2$

5) $(4)2^4$

6) 4 times y to the sixth power

7) 2 times 2 to the second power

8) $\left(\frac{2}{3}\right)^2$

9) $2x$ times y squared

10) $m^3 n^4$

11) x^0

12) $2 \cdot 5^0$

Exercises 2   Solve each expression using the order of operations.

1) $5 \cdot 5 \div 5 - 1$

2) $4 \cdot (4 + 4)$

3) $3[8 - (8 \div 2)] \div 3$

4) $4 \cdot (4 - 2)^2 \div 4 + 4$

5) $3 + 2 \cdot 3^2 \div 2 - 1$

6) $6 \div 2 \cdot 3 - 4 \div 2^2$

7) $2[(15 - 5) \div 5] + (8 - 4) \times 4$

8) $2^4 \div (2 + 2) + 2$

9) $2[2 \times (6 \div 3)] - 6 \div 3$

10) $2^3 \cdot \frac{(10 - 6)}{4} - 2$

11) $(4 + 2 \cdot 3^2) \div 2$

12) $(6 \div 3) \cdot 9 - 4 \div 2^2$

Exercises 3 Name the number set(s) that each value belongs to.

1) $5 \cdot (-2)^3$

2) $-\sqrt{3}$

3) $\dfrac{\sqrt{3}}{\sqrt{3}}$

4) $1\dfrac{4}{5}$

5) $\sqrt{3}$

6) $\left\{-8, -\dfrac{8}{1}, 8, 8.8\right\}$

7) $\left\{3, 0, \dfrac{18}{3}, \sqrt{25}\right\}$

8) $\left\{\dfrac{1}{9}, -\dfrac{18}{2}, -1.111\cdots, \sqrt{9}\right\}$

9) 4^0

10) $\sqrt{81}$

Exercises 4 Find the value of the variable in each equation.

1) If $x = 3$, find the value of $2^2 \div (12 \div x)$.

2) If $c = 4$, find the value of $\dfrac{(c)(4 \div 2) - 2}{2}$.

3) If $w = 8$, find the value of $2 \times w - 2$.

4) If $k = 4$, find the value of $(2)[(k \div 2) + 1]$.

5) If $x + 1 = 4$, find the value of $2^2 + 15 \div x$.

6) If $1 - x = -5$, find the value of $(2)[(x - 2) + 1]$.

7) If $a + 2 = 0$, find the value of $4 \times a - 2$.

8) If $k = 2$, find the value of $2[(k \div 2) - 1]$.

9) If $x - 1 = 5$, find the value of $(3^2 + 6) \div x$.

10) If $2z = 2^2$, find the value of $8 \times z - 2 \div 2$.

Exercises 5 Evaluate each expression.

1) $\dfrac{(-2)^3 + (-4)^2}{2}$

2) $\dfrac{2^3 + (-2)(4^2)}{2}$

3) $\dfrac{4^3 - 2^2}{2 \cdot (-10)}$

4) $\dfrac{-2(2^3 + 4^2)}{2 \cdot 5^2}$

5) $\dfrac{3^2 + (3)(4^2)}{3 + 4^2}$

6) $(6 \div 2) - [(4-2)(4 \div 2)^2]$

7) $[(15+5) \div (-5)]^2 - (8+4)$

8) $[(-2)^4 - 2] \div [(2^2 \times 2) - 1]$

9) $[2(5-3) \div 2](4) + 4$

10) $(2+2)^2 \div (-4^2)$

11) $\dfrac{(-2)^3}{2 \cdot 2(2-3)}$

12) $\dfrac{2 + (-4)^2}{2 \cdot 2}$

Exercises 6 Find the value of each expression.

1) If $x = -0.3$, find the value of $8x - (3 \div x)$.

2) If $x = 2$, find the value of $2(x \times 4) - 2(x \div 4)$.

3) If $x = -3$, find the value of $-2(x)^2 - (6 \div x)$.

4) If $2(x - 0.7) = 0.2$, find the value of $3.5 - 2 \div x$.

5) If $x \div 6 = \frac{1}{4}$, find the value of $2(x \div 6)$.

6) If $(x + 2) = -2$, find the value of $(2 \times x) + (2 \div x)$.

1. If $k = 6$, what is the value of $27 - 2(k + 6)$?

    A. 1   **B.** 2
    C. 3   **D.** 4

2. If $z = 2$, what is the value of $(3 - z) + 1$?

    A. 1   **B.** 2
    C. 3   **D.** 4

3. Given the quotient of 18 and x increased by 8, what is the value if $x = 9$?

    A. 10   **B.** 12
    C. 13   **D.** 14

4. Given the product of 2 and K is decreased by 2, what is the value if $K = 14$?

    A. 20   **B.** 22
    C. 24   **D.** 26

5. Given the sum of 1 and N increased by 5, what is the value if $x = 2$?

    A. 4   **B.** 6
    C. 8   **D.** 10

6. Given the difference between x and 4 divided by 2, what is the value if $3x = 6$?

    A. 2   **B.** 1
    C. −2   **D.** −1

7. Which of the following sets does not include $\sqrt{1}$?

    A. Natural number   **B.** Whole number
    C. Integer   **D.** Irrational number

8. Which of the following sets does $\dfrac{1}{\sqrt{4}}$ fall under?

    A. Natural number   **B.** Rational number
    C. Integer   **D.** Irrational number

9. Which of the following is a whole number?

    A. −2   **B.** 0

    C. 0.8   **D.** $\dfrac{1}{2}$

10. Which of the following is not a rational number?

    A. −0.5   **B.** $\sqrt{100}$

    C. 0.121   **D.** $\sqrt{39}$

11. What is the value of $(7 - 1) \div [1 + (6 - 4)] \times 4$?

    A. 6   **B.** 8

    C. 18   **D.** $\dfrac{2}{3}$

12. What is the value of $[(7 - 1) \div 3] + (8 - 4) \times 4$?

    A. 6   **B.** 8

    C. 18   **D.** $\dfrac{2}{3}$

13. What is the value of $2^2 + 2^{2}$?

    A. 4   **B.** $2 \cdot 3^2$

    C. $2 \cdot 2^2$ **D.** $2^2 \cdot 2^2$

14. What is the value of $2(2^2 + 2^2)$?

    A. 4   **B.** $2 \cdot 3^2$

    C. $2 \cdot 2^2$ **D.** $2^2 \cdot 2^2$

15. What is the value of $[(-2)^3 + (-3)^2]$?

    A. −1   **B.** 1

    C. −2   **D.** 2

16. If $-x = 3$, what is the value of $\dfrac{2}{3}(9 - 3x) \times 3$?

    A. 0   **B.** 36

    C. 18   **D.** 54

17. If $2x = 6$, what is the value of $27 - \frac{2}{3}(x + 6)$?

    A. 20   **B.** 21
    C. 19   **D.** 15

18. If $2z = 2$, what is the value of $(3 - z) + \frac{2}{3}$?

    A. $3\frac{2}{3}$ **B.** $-\frac{1}{3}$
    C. $2\frac{2}{3}$ **D.** $\frac{2}{3}$

19. Given the quotient of 8 and x to the third power increased by 8, what is the value if $x = 2$?

    A. 7   **B.** 9
    C. −7   **D.** 8

20. Given the product of 2 and K is decreased by 2 to the third power, what is the value if $K = 4$?

    A. 4   **B.** 0
    C. 6   **D.** 2

21. Given the product of 2 and x to the second power, what is the value if $x = 2$?

    A. 6   **B.** 8
    C. 10   **D.** 12

22. Given the difference between x and 2 divided by 2, what is the value if $x = 4$?

    A. 1   **B.** 2
    C. 3   **D.** 4

23. Given the difference of 6 to the third power and 3 times x, what is the value if $x = 2$?

    A. 43   **B.** 16
    C. 2   **D.** 15

24. If $z = 7$, what is the value of $(2^2 - 1)(2z + 6)$?

    A. 24   **B.** 35
    C. 32   **D.** 19

CHAPTER 2
Real Numbers

You will learn about real numbers in order to compute addition, subtraction, multiplication, and division problems expressed with two negative numbers. Absolute value, properties of equalities, and properties of real numbers are also included.

1. Adding, Subtracting, Multiplying, and Dividing Integers

* Table 1. Operations can affect your answer.

In addition:	
$(-) + (-) = -$	$(-) + (+) = +$ or $-$
$(+) + (+) = +$	$(+) + (-) = +$ or $-$
In subtraction:	
$(-) - (+) = -$	$(-) - (-) = +$ or $-$
$(+) - (-) = +$	$(+) - (+) = +$ or $-$
In multiplication:	
$(-) \times (-) = +$	$(-) \times (+) = -$
$(+) \times (+) = +$	$(+) \times (-) = -$
In division:	
$(-) \div (-) = +$	$(-) \div (+) = -$
$(+) \div (+) = +$	$(+) \div (-) = -$

2–1. Find the value of 2 + (-3).

SOLUTION

In this expression, one number is positive and the other is negative. Using a number line, start at "2", and then move 3 places to the left.

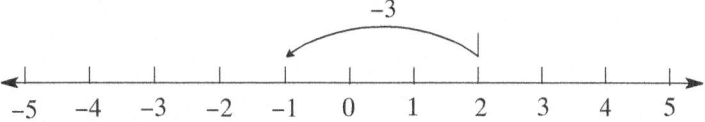

So the value in the expression is −1. Table 1 in the addition shown that is (+) + (−) = − or +. In other words, the answer takes the sign of the higher number.

Quick Exercises 1   Solve each expression

1) $(-2)+(-3)$

2) $(6)+(-7)$

3) $(-7)+8$

4) $2+3$

2–2. Find the value of $(-3)-(-2)$.

> **SOLUTION**
>
> In this expression, both numbers are negative. Start at -3 on the number line and as $-(-2) = +2$, moves 2 places to the right.
>
>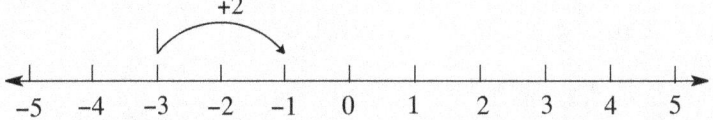
>
> So the value in the expression is -1. Table 1 shows that is $(-)-(-) = +$ or $-$.
> $-(-)$ are like signs so they become a positive sign. Therefore, the answer takes the sign of the higher number.

Quick Exercises 2   Solve each expression.

1) $(2)-(-3)$

2) $(-2)-(3)$

3) $2-3$

4) $(-2)-(-3)$

2–3. Find the value of $4 \times (-5)$.

> **SOLUTION**
>
> In the expression, one number is positive and the other is negative. Table 1 shows that $(+) \times (-) = -$. So the value of the expression is -20.
> For multiplication in an expression, if both numbers are negative, then the answer is positive. If both numbers are positive, then the answer is positive. And if one of the numbers is negative, then the answer will always be negative.

Quick Exercises 3   Find the value of each expression.

1) $6 \times (-3)^2$

2) $(-2) \times 4$

3) $(-5) \times (-2)^3$

4) 15×6

2–4. Find the value of $(-36) \div 6$.

> **SOLUTION**
>
> In the expression, one number is negative and the other number is positive. Table 1 shows that is $(-) \div (+) = -$. So, the value of the expression is -6.

Quick Exercises 4   Find the value of each expression.

1) $(-8) \div (-2)^2$

2) $8 \div (-2)^3$

3) $(-52) \div 4$

4) $78 \div 6$

Exercises 1 Find the sum of each expression.

1) $(-6) + (-7)$

2) $(-3) + (-5)$

3) $(-2)^3 + (-19)$

4) $(-32) + 13$

5) $(-61) + (-5)^2$

6) $(-7) + (-45)$

7) $(-\$26) + (-\$77)$

8) $(-\$93) + (-\$65)$

9) $(-\$60) + \$127$

10) $(-\$273) + (-\$495)$

Exercises 2 Find the difference of each expression.

1) $(-5) - (-6)$

2) $(-2) - (-3)^2$

3) $15 - (-3)^3$

4) $(-22) - 12$

5) $(-73) - (-29)$

6) $(-16) - (-36)$

7) $(-\$18) - (-\$84)$

8) $(-\$33) - (-\$65)$

9) $(-\$37) - \$29$

10) $(-\$525) - (-\$528)$

11) $(-183) - (-826)$

12) $(-901) - (-613)$

13) $(-54) - \Delta = 624$

14) $\Delta - (-243) = -284$

Exercises 3 Find the product of each expression.

1) $3 \times (-3) \times 3$

2) $(-7) \times 4$

3) $(-5) \times (-4) \times (-2)$

4) $(-5) \times (-6)$

5) $10 \times (-5) \times (-2)$

6) $\$62 \times (-8)$

7) $\$20 \times 12$

8) $-16 \times \Delta = 144$

Exercises 4 Find the quotient of each expression.

1) $(-36) \div (-3)^2$

2) $72 \div (-2)^2$

3) $(-85) \div 4$

4) $(-24) \div (-6)$

5) $(-60) \div (-3)$

6) $\$42 \div (-2)^3$

7) $\$70 \div 2$

8) $144 \div \Delta = -6$

9) $84 \div \Delta = -7$

10) $\Delta \div (-5) = -12$

Exercises 5 Find the value of Δ.

1) $-28 \div \Delta = -7$

2) $\Delta \div (-1.5) = -5$

3) $-56 \times \Delta = 672$

4) $\Delta \times (-4) = 52$

5) $49 \div \Delta = -(-7)^2$

6) $\Delta \div (-12) = -5$

7) $\$70 \div \Delta = \$17.50$

8) $(-5)^3 \div \Delta = -25$

9) $276 \div \Delta = -12$

10) $\Delta \div (-18) = 10$

2. Adding, Subtracting, Multiplying, and Dividing Rational Numbers

2–5. Adding decimals.

$$3.82 + 6.3$$

SOLUTION

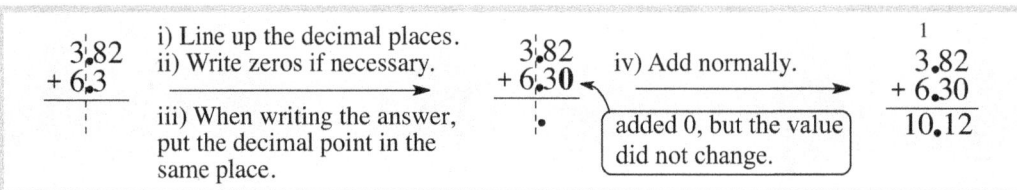

Quick Exercises 5   Find the value of each expression.

1) $9.24 + (-6.7)$

2) $\$61.39 + \$22.31$

3) $(-8) + (-4.23)$

4) $35 + \Delta = -59.43$

2–6. Subtract the decimals.

$$3 - 0.62$$

SOLUTION

$$
\begin{array}{r}
3 \\
- \ 0.62 \\
\hline
\end{array}
$$
i) Line up the decimal places
ii) Write zeros if necessary.
iii) When writing the answer, put the decimal point in the same place.

$$
\begin{array}{r}
3.00 \\
- \ 0.62 \\
\hline
. \ \ \ \\
\end{array}
$$
iv) Subtract normally.
added .00, but the value did not change.

$$
\begin{array}{r}
^{9} \\
^{2 \ \cancel{\cancel{3}} 10} \\
3.00 \\
- \ 0.62 \\
\hline
2.38 \\
\end{array}
$$

Quick Exercises 6   Find the value of each expression.

1) $2 - 0.89$

2) $(-9.4) - (-1.8)$

3) $\$89.25 - \$17.85$

4) $\Delta - (-3) = 4.53$

2-7. What is the absolute value?

a)   The absolute value is the distance of a number away from zero in any direction. For example, $|5|$ is 5 units away from zero as shown the graph below, so $|5| = 5$.

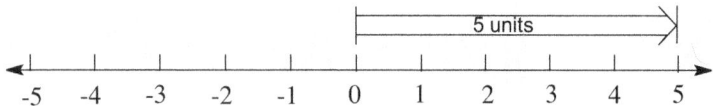

b) $|-3|$ is 3 units away from zero as shown in the graph below, so $|-3| = 3$.

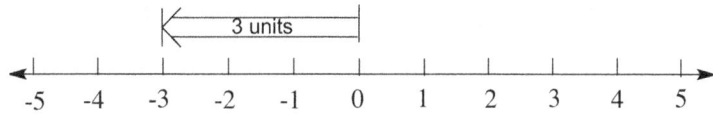

2-8. Find the value of $|-3| - 5$.

SOLUTION

The absolute value of $|-3|$ is 3.
$$|-3| - 5 = 3 - 5$$
$$= -2$$

Quick Exercises 7   Find the value of each expression.

1) $|2[(-2) + 1]| - 1$

2) $|-(2 - 1)| - 2$

3) $1 + |-2 + 1|$

4) $(-3) + |(-1) + (-2)|$

2-9. Find the value of the expression.

$$(-6) + 1\frac{1}{2}$$

SOLUTION

When the fractions have mixed numbers with a whole number, first convert it to an improper fraction.

i)

$$(-6) + \left(1\frac{1}{2}\right) = \left(-\frac{12}{2}\right) + \frac{3}{2}$$

i)

* Turn them into fractions with like denominators so that the expression will be easier to calculate.

$$(-6) = \left\{ \frac{(-6)}{1} = \frac{(-6) \times 2}{1 \times 2} = \frac{(-12)}{2} \right\}$$

Equivalent fractions

i) Change the mixed numbers to improper fractions.*

ii)

$$(-\frac{12}{2}) + \frac{3}{2} = \overbrace{\frac{(-12) + 3}{2}} = \frac{-9}{2} \quad \text{or} \quad -4\frac{1}{2}$$

iii)

like denominators;
the same denominators

ii) Add the fractions.
iii) Simplify if necessary.

* Improper fraction: A fraction with a numerator that is greater than or equal to the denominator in the fraction. For example; $\frac{3}{1}$, $\frac{6}{5}$, or $\frac{4}{4}$

Quick Exercises 8   Find the value of each expression.

1) $1 + \frac{2}{3}$

2) $\frac{5}{2} + 0.25$

3) $(-2) + 1\frac{3}{4}$

4) $(-\frac{2}{5}) + (-3)$

2-10. Find the value of the expression.

$$2\frac{1}{8} - (-3)$$

SOLUTION

If the mixed fractions have unlike denominators, then turn them into fractions with like denominators. Change the mixed numbers to improper fractions, subtract, and simplify if necessary.

i)

$$2\frac{1}{8} - (-3) = \frac{17}{8} - \frac{(-24)}{8}$$

i)

* Make the like denominators that are easy to calculate with other fraction.

$$(-3) = \left\{ \frac{(-3)}{1} = \frac{(-3) \times 8}{1 \times 8} = \frac{(-24)}{8} \right\}$$

equivalent fractions

i) Change the mixed numbers to improper fractions.

$$\frac{17}{8} - \frac{(-24)}{8} = \overbrace{\frac{17-(-24)}{8}}^{\text{ii)*}} = \frac{41}{8} \quad \text{or} \quad 5\frac{1}{8}$$

ii) iii)

like denominators;
the same denominators

ii) Subtract the fractions with like denominators.   * $17 - (-24) = 17 + 24$
iii) Simplify if necessary.

Quick Exercises 9   Find the value of each expression.

1) $(-1) - \dfrac{1}{2}$

2) $1\dfrac{1}{3} - 1$

3) $2 - 1\dfrac{7}{8}$

4) $\left(-\dfrac{2}{5}\right) - (-3)$

Exercises 6 Add the decimals.

1) $(-3.24) + 6.47$

2) $\$61.39 + \$22.31$

3) $(-3) + 4.23$

4) $(-35) + 9.43$

5) $0.253 + (-1)$

6) $-24.073 + 18.059$

7) $\$25.65 + (-\$78.08)$

8) $(-0.606) + (-0.597)$

9) $7.09 + 9.96$

10) $-\$38.57 + \$75.99$

Exercises 7 Subtract the decimals.

1) $63.84 – $17.89

2) $79.26 – 47.99$

3) $-12.366 – (-7.07)$

4) $21.05 – $13.29

5) $(-11.68) – 27.07$

6) $41.72 – 13.37$

7) $0.6 – 0.49$

8) $(-0.50) – (-0.26)$

9) $-7.36 – (-3.07)$

10) $4.05 – $7.29

11) $7.62 – 2.77$

12) $41 – 13.43$

Exercises 8 Find each value of each expression.

1) $|2 - 1| - 1$

2) $6 + |-2 - 4|$

3) $|3[(-4) + 2]| - 4$

4) $|-2(5 - 1)| - 5$

5) $5 + |-8 + 6|$

6) $(-6) + |(-2) + (-4)|$

7) $|(-2) - (-1)| - 1$

8) $6.2 + |-2.5 - 4.9|$

Exercises 9 Find each value of each expression.

1) $-|8-(-2)|$

2) $0.75 + |-2.5|$

3) $|-[(-4) + 2]| - 4$

4) $|-(5 - 1)| - 5$

5) $(-5) + |-8 + 6|$

6) $6 + |(-2) + (-4)|$

7) $8 - |(1 - 5)|$

8) $|-7 - 2| - |-2(2 - 5)|$

9) $|-(9 - 2)| + |-2(2 - 4)|$

10) $|-2(5 - 14)| - |-14|$

Exercises 10 Find the value of each expression.

1) If $x = -1$, find the value of $|2x - 1| - 1$.

2) If $x = 2$, find the value of $|-2 - 4x| - 6$.

3) If $x = -4$, find the value of $\left|\frac{1}{4}(x + 2)\right| - 4$.

4) If $x = 2$, find the value of $|-(5 - x)| - 5$.

5) If $2x = 1$, find the value of $|-2x + 6| + 5$.

6) If $2x = -6$, find the value of $|-2(x + 2)| - 5$.

7) If $x - 3 = 0$, find the value of $-2 \times |(-x + 2)|$.

8) If $x - 2^2 = 0$, find the value of $-2 \times |-2(5 - x)|$.

Exercises 11 Add the fractions and simplify if necessary.

1) $2 + \dfrac{2}{5}$

2) $\left(-\dfrac{1}{4}\right) + 1$

3) $\dfrac{2}{3} + (-3)$

4) $2 + 1\dfrac{5}{8}$

5) $\left(-1\dfrac{1}{8}\right) + 6$

6) $|(-2) + (-4)| + 1\dfrac{3}{4}$

7) $2\dfrac{1}{6} + 2.5$

8) $(-4.75) + |(-2) - 4|$

9) $\left(-\dfrac{5}{9}\right) - |(-1) + (-1)|$

10) $1\dfrac{2}{3} + 4\dfrac{1}{6}$

Exercises 12 Subtract the fractions and simplify if necessary.

1) $(-2)^2 - \dfrac{2}{3}$

2) $2^2 - 2\dfrac{6}{7}$

3) $2\dfrac{1}{4} - (-1)^3$

4) $2\dfrac{1}{5} - 2$

5) $1\dfrac{3}{9} - |(-2) + 1|$

6) $1\dfrac{1}{3} - (-1)^2$

7) $|(-2) - (-3)| - \left(-\dfrac{1}{5}\right)$

8) $\left(-3\dfrac{1}{6}\right) - 2\dfrac{1}{2}$

2-11. Multiplying decimals.

$$2.9 \times 7$$

SOLUTION

$$\begin{array}{r} 2.9 \\ \times\ \ 7 \\ \hline \end{array} \longleftarrow \textbf{1 decimal place}$$

i) $\quad \begin{array}{r} \overset{6}{2.9} \\ \times\ \ 7 \\ \hline 203 \end{array}$

ii) $\quad \begin{array}{r} 2.9 \\ \times\ \ 7 \\ \hline 20.3 \end{array} \Big\}\ \textbf{1 decimal place}$

$\qquad\qquad$ 1 decimal place

i) Multiply and regroup.

ii) After finding the product, put the decimal point in the answer-
it will have as many decimals places as the two numbers being multiplied.

So, the answer is 20.3.

Quick Exercises 10   Find the value of each expression.

1) $(-1.2) \times (-3)^2$

2) $(-8) \times (-2.4)$

3) $2 \times (-2)^3$

4) 4.5×3

2-12. Dividing decimals.

$$7.2 \div 3$$

SOLUTION

a) First, line up the decimal point with the dividend.
b) Divide them as if they are whole numbers.

$3\overline{)7.2}$

i) Line up the decimal
point with the dividend. $\longrightarrow$ $3\overline{)7.2}^{\ .}$

ii) Divide as if the dividend is a
whole number. $\longrightarrow$

$$\begin{array}{r} 2.4 \\ 3\overline{)7.2} \\ -6\ \downarrow \\ \hline 12 \\ -12 \\ \hline 0 \end{array}$$

divisor

$3 \times 2 = 6$

$3 \times 4 = 12$

Quick Exercises 11   Find the value of each expression.

1) $(-1.6) \div (-2)^2$

2) $8.2 \div 2$

3) $(-5.4) \div 3$

4) $18 \div (-2.5)$

2-13. Multiply the fractions.

$$3\frac{1}{2} \times \frac{2}{5}$$

SOLUTION

If they are unlike fractions, then i) change the mixed numbers to improper fractions, ii) before multiplying fractions, simplify using the GCF, iii) multiply and simplify if necessary.

$$3\frac{1}{2} \times \frac{2}{5} \quad \xrightarrow{\text{i) Change the mixed numbers to improper fractions.}} \quad \frac{7}{2} \times \frac{2}{5}$$

$$\frac{7}{2} \times \frac{2}{5} \quad \xrightarrow{\text{ii) Before multiplying fractions, simplify with the GCF.}} \quad \frac{7}{\cancel{2}_1} \times \frac{\cancel{2}^1}{5}$$

$$\frac{7}{\cancel{2}_1} \times \frac{\cancel{2}^1}{5} = \frac{7}{1} \times \frac{1}{5} = \frac{7}{5} \quad \text{or} \quad 1\frac{2}{5}$$

iii) Multiply and simplify if necessary.

So, the product is $1\frac{2}{5}$.

* Improper fractions: A fraction where the numerator is greater than or equal to the denominator. For example; $3 = \frac{3}{1}$, $1\frac{1}{5} = \frac{6}{5}$, or $1 = \frac{4}{4}$

Quick Exercises 12   Find the value of each expression.

1) $(-3)^3 \times \frac{2}{3}$

2) $(-2)^2 \times (-1\frac{2}{5})$

2-14. Dividing fractions.

$$2\frac{3}{5} \div 3$$

SOLUTION

If they are unlike fractions, then i) change the mixed numbers to improper fractions, ii) flip the numerator and denominator of the divisor, iii) change the operation sign, iv) multiply and simplify if necessary.

ii) Flip (reciprocal)* of fraction.

$2\dfrac{3}{5} \div 3$ i) Change the mixed numbers to improper fractions. $\boxed{3 = \dfrac{3}{1}}$ $\longrightarrow$ $\dfrac{13}{5} \div \dfrac{3}{1} = \dfrac{13}{5} \times \dfrac{1}{3}$

iii) Change the operation sign.

$\dfrac{13}{5} \times \dfrac{1}{3}$ iv) Multiply. $\longrightarrow$ $\dfrac{13}{15}$

So, the quotient of $3\dfrac{3}{5} \div 3$ is $\dfrac{13}{15}$.

* Reciprocals: two numbers are reciprocals if their product is 1. For example;

$$\dfrac{1}{2} \times \dfrac{2}{1} = 1 \qquad \dfrac{5}{7} \times \dfrac{7}{5} = 1 \qquad \dfrac{9}{10} \times \dfrac{10}{9} = 1$$

Quick Exercises 13   Find the value of each expression.

1) $(-1)^2 \div \dfrac{1}{2}$

2) $\dfrac{2}{3} \div (-2)^2$

2–15. Mixed numbers.

$$(-2)^4 \text{ and } (-2)^5$$

SOLUTION

$(-2)^4$ and $(-2)^5$ — exponents, bases

In an exponential expression, 2 is the base. 4 and 5 are the exponents. The sign of the base affects the exponent.

If the exponent is an even number, then the sign will be positive (+). If the exponent is an odd number, then the sign will be negative (–). So $(-2)^4$ will be positive because an even number of negatives multiplied is positive and $(-2)^5$ will be negative because an odd number of positives multiplied is odd.

$(-2)^0 = 1$
$(-2)^1 = (-2) = -2$
$(-2)^2 = (-2) \times (-2) = 4$
$(-2)^3 = (-2) \times (-2) \times (-2) = -8$
$(-2)^4 = (-2) \times (-2) \times (-2) \times (-2) = 16$
$(-2)^5 = (-2) \times (-2) \times (-2) \times (-2) \times (-2) = -32$

3. Real Numbers

2–16. Find the square roots.

$$\sqrt{4} \text{ and } \sqrt{6}$$

SOLUTION

Symbol : $\sqrt{}$ is called a radical.
There are two kinds of square roots. Perfect squares are the squares of whole numbers,
such as $\sqrt{4} = 2$ and $\sqrt{16} = 4$.

Square roots can be expressed as irrational numbers such as $\sqrt{5} = 2.236067\cdots$ and $\sqrt{6} =$
$2.44948\cdots$. So $\sqrt{5}$ and $\sqrt{6}$ are real numbers that are rational and irrational numbers.

Quick Exercises 14   Find the value of each expression.

1) $-\sqrt{28}$

2) $\sqrt{-(1)(9)}$

3) $\pm\sqrt{49}$

4) $\dfrac{\sqrt{4}}{\sqrt{25}}$

Exercises 13 Multiply each expression.

1) 91×7

2) $(-5) \times (-2)$

3) 0.7×2

4) $\$8.63 \times \$9.08$

5) $62 \times (-7)$

6) $(-1.2) \times 9$

7) 4×1.5

8) $\$8.63 \times 2$

9) $\$0.25 \times 7$

10) $(-6.2) \times (-3)$

11) $(-0.75) \times 4$

12) $\$25 \times 5$

13) $(-1.6) \times (-9)$

14) $(-5) \times (-0.5)$

Exercises 14 Divide each expression.

1) $(-5.6) \div (-4)$

2) $(-9.6) \div 2$

3) $96 \div (-8)$

4) $0.35 \div 5$

5) $1.8 \div 3$

6) $(-1.2) \div 6$

7) $75 \div 3$

8) $0.75 \div 3$

9) $(-40.8) \div 6$

10) $2.8 \div 5$

11) $0.04 \div 2$

12) $(-32) \div (-8)$

13) $1.6 \div 0.4$

14) $1.26 \div 6$

Exercises 15 Multiply each expression.

1) $(-0.26)(-0.5)$

2) $(4)(\frac{6}{7})$

3) $(-\frac{1}{3})(\frac{2}{5})$

4) $(\frac{3}{8})(2)$

5) $(-\frac{4}{7})(\frac{1}{7})$

6) $(-\frac{5}{12})(-\frac{2}{3})$

7) $(\frac{2}{5})(\frac{5}{8})$

8) $(-\frac{3}{10})(-\frac{2}{3})$

9) $(-\frac{5}{7})(\frac{2}{5})$

10) $(-\frac{3}{4})(-\frac{2}{9})$

11) $(3\frac{2}{9})(2\frac{5}{8})$

12) $(-\frac{3}{5})(-1\frac{2}{3})$

Exercises 16 Divide each expression.

1) $\dfrac{5}{7} \div 5$

2) $1\dfrac{4}{9} \div 2\dfrac{2}{3}$

3) $\dfrac{3}{14} \div 3\dfrac{2}{2}$

4) $2\dfrac{3}{8} \div 1\dfrac{3}{16}$

5) $2\dfrac{1}{2} \div 3\dfrac{1}{3}$

6) $1\dfrac{2}{3} \div 5$

7) $3\dfrac{1}{2} \div 1\dfrac{1}{6}$

8) $2\dfrac{2}{6} \div 4\dfrac{2}{7}$

9) $\dfrac{7}{5} \div 2$

10) $1\dfrac{1}{2} \div 2\dfrac{1}{2}$

11) $\dfrac{3}{7} \div 1\dfrac{2}{4}$

12) $2\dfrac{1}{6} \div 1\dfrac{3}{8}$

Exercises 17 Simplify without using a calculator.

1) $\sqrt{98}$

2) $-\sqrt{27}$

3) $\sqrt{-(4)(9)}$

4) $\pm\sqrt{50}$

5) $-\sqrt{32}$

6) $\sqrt{-0.49}$

7) $\sqrt{\dfrac{4}{25}}$

8) $\dfrac{\sqrt{4}}{\sqrt{2}}$

9) $\sqrt{108}$

10) $-\sqrt{8}$

11) $\sqrt{(5)(20)}$

12) $\pm\sqrt{48}$

1. What is the sum of the expression below?
$$6 + (-3)$$

    A. 3   **B.** 6

    C. −3   **D.** 9

2. What is the sum of the expression below?
$$(-2) + (-8)$$

    A. −4   **B.** −6

    C. −8   **D.** −10

3. What is the difference of the expression below?
$$(-7) - 5$$

    A. −2   **B.** −8

    C. −10   **D.** −12

4. At Halloween, Mr. Myers gives away 7 pieces of candy to each trick-or-treater. If 57 trick-or-treaters came to his door, how many pieces of candy did he give away?

    A. 420   **B.** 50

    C. 64   **D.** 399

5. A bakery has 45 bags of bagels. Each bag contains 12 bagels. How many bagels does the bakery have?

    A. 3.75   **B.** 495

    C. 3.5   **D.** 540

6. A tray can hold 65 eggs. If there are 1,495 eggs, how many trays will be needed?

    A. 1,560   **B.** 1,430

    C. 97,175   **D.** 23

7. At an art exhibition, nails are needed in order to hang up the paintings. If a painting needs 14 nails to stay on the wall and there are 168 paintings, how many nails are needed in total?

    A. 12   **B.** 182

    C. 154   **D.** 2,352

8. What is the product of the expression below?
$$250 \times (-7)$$

    A. −2,100   **B.** −1,750
    C. −2,500   **D.** −257

9. What is the difference of the expression below?
$$10 - (-17)$$

    A. −27   **B.** −7
    C. 27   **D.** 7

10. What is the quotient of the expression below?
$$(-63) \div (-7)$$

    A. −70   **B.** −9
    C. −56   **D.** 9

11. What is the value of Δ for the expression below?
$$(-24) \div \Delta = -12$$

    A. 2   **B.** −2
    C. 288   **D.** −288

12. What is the value of Δ for the expression below?
$$19 - \Delta = 20$$

    A. −1   **B.** 39
    C. 1   **D.** 20

13. What is the value of $(-10) + |(-1) + (-2)|$?

    A. −10   **B.** −13
    C. −7   **D.** 7

14. What is the value of $|(-5) - (-7)| - (-3)$?

    A. 1   **B.** 3
    C. 6   **D.** 0

15. What is the value of $(-1\frac{1}{6}) \times |(-8) - (-2)|$?

    A. −7   **B.** −42
    C. 42   **D.** 7

16. If $x = 2$ and $y = -4$, what is the value of $x + y$?

    A. 6   **B.** −2
    C. 4   **D.** 2

17. If $x = 2$ and $y = -1$, what is the value of $-\frac{2}{3}[(-x) + y]$?

    A. 2   **B.** −2
    C. $1\frac{1}{3}$ **D.** $-1\frac{1}{3}$

18. If $x = 2$ and $y = 2$, what is the value of $(-x) + (-y)$?

    A. 4   **B.** 8
    C. −4   **D.** 0

19. If $(x - 3)^2 = 25$ and $y = 4$, what is the value of $\frac{1}{4}[x \times (-y)]$?

    A. 8   **B.** $1\frac{1}{4}$
    C. −8   **D.** −6

20. If $2x = 6$ and $y = (-1)^2$, what is the value of $(-x) - y$?

    A. 4   **B.** −2
    C. −4   **D.** 2

21. If $x = -1\frac{2}{3}$ and $y = -1\frac{1}{2}$, what is the value of $(x)(y)$?

    A. $2\frac{1}{2}$ **B.** $-2\frac{1}{2}$

    C. $1\frac{1}{3}$ **D.** $-1\frac{1}{3}$

22. If $x = 5$ and $y = -1\frac{1}{4}$, what is the value of $(-x) \div y$?

    A. −5   **B.** 5
    C. 4   **D.** −4

22. If $x = -2$ and $y = -5$, what is the value of $\dfrac{2[(-x) - y]}{3}$?

     A. $4\dfrac{2}{3}$ **B.** 2

     C. $-4\dfrac{2}{3}$ **D.** $\dfrac{2}{3}$

23. If $x^2 = 5$ and $y^3 = -1$, what is the value of $-x^2 \times y^3$?

     A. 5 **B.** -5

     C. 25 **D.** -25

24. If $x = 2$, what is the value of $(-x)\sqrt{1}$?

     A. 2 **B.** -2

     C. ± 2 **D.** $-\sqrt{-2}$

25. If $x = |-1|$ and $y = (2)^2$, what is the value of $(x)\sqrt{y}$?

     A. 2 **B.** -2

     C. ± 2 **D.** $-\sqrt{2}$

25. If $x = (-2)^2$ and $\sqrt{y} = 3$, what is the value of $(x)\sqrt{y}$?

     A. 12 **B.** 48

     C. $\pm 16\sqrt{3}$ **D.** $4\sqrt{3}$

26. If $x^2 = 2x$ and $2\sqrt{y} = 16$, what is the value of $(x) + \sqrt{y}$?

     A. $2 + 2\sqrt{2}$ **B.** 18

     C. 10 **D.** $4\sqrt{2}$

27. Which of the following statements are true?

     A. Square roots are always irrational.
     B. Square roots are not always real numbers.
     C. Square roots are not always integers.
     D. $\sqrt{4} = \pm 2$

28. Which of the following is true?

A. $-\dfrac{1}{3} > \sqrt{4}$ B. $-\dfrac{4}{5} > \sqrt{4}$

C. $\sqrt{4} < \dfrac{1}{4}$ D. $\dfrac{1}{3} < \sqrt{4}$

29. Which of the following is true?

A. $1 > \sqrt{1}$ B. $1\dfrac{1}{3} > \sqrt{3}$

C. $\sqrt{1} < \dfrac{1}{\sqrt{1}}$ D. $1\dfrac{1}{2} < \sqrt{4}$

30. Which of the following is true?

A. $1 > |-1|$ B. $-|1| < 1^{0}$

C. $\sqrt{1} < |-1|$ D. $(-1)^{2} = -1^{2}$

31. Which of the following is true?

A. $3 > \sqrt{12}$ B. $9 > \sqrt{83}$

C. $6 < \sqrt{32}$ D. $10 < \sqrt{101}$

32. Which of the following numbers is an integer?

A. 0.5 B. -4

C. $\dfrac{2}{6}$ D. $\sqrt{5}$

33. Which of the following correctly lists the numbers below from least to greatest?

$$2.01 , \ \sqrt{5}, \ \sqrt{4}, \ 0$$

A. $\sqrt{5} > \sqrt{4} > 2.01 > 0$ B. $\sqrt{5} < 2.01 < \sqrt{4} < 0$

C. $0 < \sqrt{4} < 2.01 < \sqrt{5}$ D. $0 < \sqrt{4} < \sqrt{5} < 2.01$

34. Which of the following correctly lists the numbers below from least to greatest?

$$\sqrt{2} , \ 1.\overline{25}, \ \sqrt{\dfrac{25}{16}}, \ \dfrac{\sqrt{4}}{3}$$

A. $\sqrt{\dfrac{25}{16}} > \sqrt{2} > 1.\overline{25} > \dfrac{\sqrt{4}}{3}$ B. $1.\overline{25} < \sqrt{2} < \dfrac{\sqrt{4}}{3} < \sqrt{\dfrac{25}{16}}$

C. $\dfrac{\sqrt{4}}{3} < \sqrt{2} < 1.\overline{25} < \sqrt{\dfrac{25}{16}}$ D. $\dfrac{\sqrt{4}}{3} < \sqrt{\dfrac{25}{16}} < 1.\overline{25} < \sqrt{2}$

4 Properties of Numbers

2–17. Name the property used for each step when solving the equation.

$$2 + 4(3 + x) + 2(5 + 1) = 7^2$$

SOLUTION

$2 + 4(3 + x) + 2(5 + 1) = 7^2$	Original equation.
$2 + 12 + 4x + 2(5 + 1) = 7^2$	Distributive Property
$2 + 12 + 4x + 2(6) = 72$	Compute the exponential

* Properties of Numbers

Associative Property of Addition: $(2 + 5) + 3 = 2 + (5 + 3)$
Associative Property of Multiplication: $[(2)(5)](3) = (2)[(5)(3)]$
Commutative Property of Addition: $2 + 5 = 5 + 2$
Commutative Property of Multiplication: $(2)(5) = (5)(2)$
Distributive Property: $2(5 + 3) = (2)(5) + (2)(3), 2(5 - 3) = (2)(5) - (2)(3)$
Reflexive Property: $2 + 5 = 2 + 5, 2 = 2$
Identity Property of Addition: 0 is the identity. $2 + 0 = 0 + 2 = 2$
Identity Property of Multiplication: 1 is the identity. $(2)(1) = (1)(2) = 2$
Substitution: If $x = 3$ and $x + 7 = 10$, then 3 can be substituted in $x + 7 = 10$ to obtain
      $3 + 7 = 10$

Quick Exercises 15   Name the property used for each step when solving the equation.

1) If $4(x - 5) = 4x - 20$

2) $2 + (3 + 7) = (2 + 3) + 7$

3) If $x(10 + 3) = 1$, then $13x = 1$.

4) $x + 9 = 9 + x$

2–18. Simplify the equation.

$$\frac{1}{6}x(6y - 3) = 2 + x - 2xy$$

SOLUTION

Use the order of operations system.
i) Parentheses, ii) Exponents, iii) Multiply or divide from left to right, iv) Add or
  subtract in order from left to right.

$\frac{1}{6}x(6y - 3) = 2 + x - 2xy$	Original equation
$xy - 0.5x = 2 + x - 2xy$	Distributive Property. $6 \times \frac{1}{6} = 1$
$xy - 0.5x + 0.5x = 2 + x - 2xy + 0.5x$	Add $0.5x$ to both sides.
$xy = 2 + 1.5x - 2xy$	Simplify.
$xy - xy = 2 + 1.5x - 2xy - xy$	Subtract xy from both sides.
$0 = 2 + 1.5x - 3xy$	Simplify.

Quick Exercises 16   Simplify the equation.

1) $4(x - 5) = 2(x + 10)$

2) $2 + \dfrac{1}{3}(3x + 7) = (1 + 3x) - 7x$

3) $2x + 3 = 13x - 1$

4) $x(2 + 9y) - 12 = 9 + 4x - 2xy$

Exercises 18   Determine the steps involved in solving each equation.

1) $2(x - 3) = 2x - 6$

2) $x + 3 = x + 3$

3) $0 + 9 = 9$

4) $x \cdot 17 = 17 \cdot x$

5) $(x)(1) = x$

6) $(a + b) + c = a + (b + c)$

7) $x = x$

8) If $x + (8 + 15)y = 13$, then $x + 23y = 13$

9) If $(1 + 2)x = 6$, then $3x = 6$

10) $x(2 + 3) = 2x + 3x$

11) $(x)(y) = (y)(x)$

12) $(1)(a) = a$

13) $2 + x = 2 + x$

14) If $(1 + 2)x = 6$, then $x + 2x = 6$

Exercises 19 Simplify each expression.

1) $-(3x - 2) = 2(x + 3)$

2) $\frac{1}{4}x(1 + 8y) = xy + 3x$

3) $4x - \frac{2}{3}(9x + 3)$

4) $x(1 + 3y) - 2xy$

5) $x - 5 + 2(-5 + x)$

6) $(1 + 3x) = \frac{2}{5}(3x + 1)$

7) $2x + 3(2x - 1)$

8) $x(2 + 9y) - 2xy$

9) $4(x - 5) - 2(x + 10)$

10) $2 + \frac{4}{5}(x + 2) = 1 + 3x$

11) $(-x) + (-3) = 2x - (-1)$

12) $x(2 + 9y) = 4(x - 2xy)$

13) $2(x + 3) - (-x - 3)$

14) $x(1 + 2y) - x$

15) $5(x - 1) - 2(x - 1)$

16) $4 - (x + 2) = 2 - 3x$

17) $(x + 3) + 2[x - (-3)]$

18) $xy(2 - y) + 4xy(-y + 2)$

1. Which of the following equations represents the Distributive Property?

   A. If $\frac{2}{9}(7 + 11) = 4$, then $\frac{2}{9}(18) = 4$ **B.** If $-2[2 + (-3)] = x$, then $2 = x$

   C. $2(x \times x \times x) = 2x \times 2x \times 2x$ **D.** $2(x \times x \times x) = 2(3x)$

2. Which of the following equations represents the Substitution Property?

   A. If $\frac{2}{9}(7 + 11) = 4$, then $\frac{2}{9}(18) = 4$ **B.** If $-2[2 + (-3)] = x$, then $2 = x$

   C. $2(x \times x \times x) = 2x \times 2x \times 2x$ **D.** $2(x \times x \times x) = 2(3x)$

3. Given $-(4x \times 3) = (-4x) \times (-3)$, what is the property that represents the equation?

   A. Distributive Property   **B.** Substitution Property
   C. Reflexive Property   **D.** Associative Property of Addition

4. Which of the following equations represents the Commutative Property of Addition?

   A. $(2 + x) + 3 = 2 + (x + 3)$ **B.** $2(x + 3) = 2x + 6$
   C. $3x + 2 = 2 + 3x$ **D.** $(2 + x) + 0 = 2 + x$

5. Given $7 + 11 = 7 + 11$, what is the property in the statement?

   A. Distributive Property   **B.** Substitute Property
   C. Reflexive Property   **D.** Associative Property of Addition

6. Which of the following is equal to $2 \times (x + 10)$?

   A. $2x + 10$ **B.** $2x \times 2 + 10$
   C. $2x + 20$ **D.** $(2 + x) \times (2 + 10)$

7. Which of the following is NOT equal to $2x(y - 5) = \frac{3}{2}[(-5x) + xy]$?

   A. $xy - 5x = 0$ **B.** $4xy - 20x = (-15x) + 3xy$
   C. $2x(y - 5) = 0$ **D.** $4x(y - 5) = 3[(-5x) + xy]$

8. Which of the following is NOT equal to $2(\frac{1}{2}x + 10x)$?

A. $21x$ **B.** $x \times 20x$

C. $x + 20x$ **D.** $2x(\frac{1}{2} + 10)$

9. Which of the following is simplified from $x(3y + 6) - 3x(2 - 2y)$?

A. $6x + 6xy$ **B.** $3xy + 6x - 6y$
C. $6xy + 6x - 6y$ **D.** $6x - 6xy$

10. Which of the following is equivalent to $2(x + 1)$?

A. $2x$ **B.** $3x$
C. $2x + 2$ **D.** $2x + 1$

11. Which of the following is simplified from $5(\frac{4}{5}x + 2x) - 2(5x + \frac{5}{2}x)$?

A. x **B.** $-x$
C. $9x$ **D.** $4\frac{7}{10}x$

12. What is the value of $52 - 2^4 \cdot 3 - 1$?

A. 107 **B.** 3
C. 72 **D.** 20

13. What is the value of $(5 - 3)^4$?

A. 625 **B.** 81
C. $5^4 - 3^4$ **D.** 16

14. Which of the following numbers is a perfect square?

A. 2 **B.** 3
C. 24 **D.** 25

CHAPTER 3
Solving Linear Equations

> You will learn how to solve linear equations with one or two variables using addition, subtraction, multiplication, and division. Problems involving rates, average speed, distance, and time are also included.

1. Solving Equations with a Variable: Addition and Subtraction

3-1. Find the value of the variable.

$$2 + x = -3$$

SOLUTION

i) An equation is considered correct when both sides of the equal sign are equal. So the left side $(2 + x)$ should have the same value as the right side of the equation (-3).

ii) Now you should find the value of x. To solve for x, you should subtract 2 from both sides.

$$2 - 2 + x = -3 - 2$$

The left side of the equation can be canceled out $(2 - 2 = 0)$. Therefore, only x remains. On the right side, 2 is subtracted from -3.　$(-3 - 2 = -5)$

$$x = -5$$

So the value of x is -5.

Check to verify x is -5.

$2 + x = -3$	Original equation
$2 + (-5) = -3$	Replace x with -5.
$-3 = -3$	

So both sides are equal. This means that $x = -5$.

Quick Exercises 1　　Solve each equation.

1) $x + 6 = 21$

2) $13 + y = 32$

3) $3.5 + x = 19.7$

4) $n + 4.65 = 12.43$

3-2. Find the value of x.

$$x - 6 = -3$$

SOLUTION

Look at the equation of $x - 6 = -3$, where x is the unknown value.

$x - 6 + \mathbf{6} = -3 + \mathbf{6}$	Add 6 to both sides.
$x = 3$	Simplify.

Check the solution:

$x - 6 = -3$	Original equation
$3 - 6 = -3$	Substitute x with 3.
$-3 = -3$	Simplify. This means the answer is correct.

Quick Exercises 2   Solve each equation

1) $6 - x = 2$

2) $2(n - 15) = 4$

3) $c - 17.8 = 26.2$

4) $5(2.52 - y) = 2.25$

3-3. Find the value of x.

$$\frac{x}{4} - 7 = 2 \text{ or } (x \div 4) - 7 = 2$$

SOLUTION

a) Look at the equation $\frac{x}{4} - 7 = 2$, where x is the unknown value.

$\frac{x}{4} - 7 = 2$	Original equation
$\frac{x}{4} - 7 + 7 = 2 + 7$	Add (7) to both sides.
$\frac{x}{4} = 9$	Simplify. $-7 + 7 = 0,\ 2 + 7 = 9$
$4 \times \frac{x}{4} = 9 \times 4$	Multiply each side by 4.
$x = 36$	Simplify. $4 \times \frac{x}{4} = x,\ \ 9 \times 4 = 36$

So, the value of x is 36.

Check the solution.

$\frac{x}{4} - 7 = 2$	Original equation
$\frac{36}{4} - 7 = 2$	Substitute x with 36.
$9 - 7 = 2$	Simplify the fraction.
$2 = 2$	This means the answer is correct.

Quick Exercises 3   Find each value of x.

1) $1 + x = \dfrac{1}{2}$

2) $4x + 1\dfrac{3}{4} = 6$

3) $4 + (x \div 2) = 10$

4) $\dfrac{2}{x} + 3 = -3$

3–4. Find the value of x.

$$2 - \dfrac{x}{5} = -2 \text{ or } 2 - (x \div 5) = -2$$

SOLUTION

Look at the equation $2 - \dfrac{x}{5} = -2$, where x is the unknown value.

$2 - \dfrac{x}{5} = -2$	Original Equation
$(2 - 2) - \dfrac{x}{5} = [(-2) - 2]$	Subtract 2 from both sides.
$-\dfrac{x}{5} = -4$	Simplify. $(2 - 2) = 0$, $[(-2) - 2] = -4$
$(-5) \times (-\dfrac{x}{5}) = (-4) \times (-5)$	Multiply each side by –5.
$x = 20$	Simplify. $(-5) \times (-\dfrac{x}{5}) = x$, $(-4) \times (-5) = 20$

So, the value of x is 20.

Check the solution.

$2 - \dfrac{x}{5} = -2$	Original equation
$2 - \dfrac{20}{5} = -2$	Substitute x with 20.
$2 - 4 = -2$	Simplify. $\dfrac{20}{5} = 4$
$-2 = -2$	The answer is correct.

Quick Exercises 4   Solve each equation

1) $y - 2 = 3\dfrac{2}{3}$

2) $(x \div 3) - 1 = -2$

3) $2 - \dfrac{x}{3} = 7$

4) $\dfrac{1}{2y} - 1 = 1$

Exercises 1 Solve each equation.

1) $3c + 5 = -28$

2) $14 + 5y = 42.5$

3) $2.8 + 2x = 7.4$

4) $5n + 1 = 6$

5) $2(c + 17) = -14$

6) $25 + 2y = 31$

7) $-2 + 8x = 2$

8) $3x + 8.25 = -4.65$

9) $-5x + 10.9 = -9.75$

10) $-4 + 8y = 28$

Exercises 2 Find the value of each expression.

1) Given $x = 5$, find the value of $7(2 - x)$.

2) Given $c = -3$, find the value of $2(4 - c)$.

3) Given $2x = -0.8$, find the value of $2.35 + x$.

4) Given $0.5y = 2.5$, find the value of $2(y + 1.5)$.

5) If $x + 2 = -1$, find the value of $7(2 - x)$.

6) If $-3 + c = 1$, find the value of $3(c - 3)$.

7) If $2x = -4$, find the value of $1 + 2x$.

8) If $0.5y = -2$, find the value of $2 - 0.5y$.

Exercises 3 Solve each equation.

1) $4c - 2.4 = 16$

2) $10 - 3y = 31$

3) $4.3 - 2x = -8.3$

4) $8.4n - 1 = 41$

5) $2(c - 1.2) = 5.6$

6) $4(-3 - y) = -16$

7) $-3 - 5x = -40$

8) $3x - 8 = -50$

9) $y - 1.8 = 2$

10) $2.4 - y = -2.4$

Exercises 4 Find the value of each expression.

1) Given $x = 2$, find the value of $2(8 - x)$.

2) Given $c = 18$, find the value of $14 - c + 8$.

3) Given $x = 0.8$, find the value of $9 + x$.

4) Given $y = 6.4$, find the value of $3(y + 1.6)$.

5) If $n = 0.5$, find the value of $8 - 4n$.

6) If $c = 11$, find the value of $2.5(c - 3)$.

7) If $x = 4$, find the value of $12 - 0.2x$.

8) If $y = 5.5$, find the value of $2(y - 2.1)$.

Exercises 5 Solve each equation.

1) c − 1.2 = 5.6 − 3c

2) −3 − 2y = −18 + 2y

3) 2(−3 − 5x) = −4x

4) 3x − 9 = −5x + 9

5) 0.5y − 2.8 = 2.5y + 5.2

6) 6.9 − x = −3.3x

7) 2(7 − x) = 6(x + 1)

8) 3(x − 1) = 5(x + 1)

9) y − 0.8 = 0.4y + 7.2

10) 3.0 − x = −0.3x

Exercises 6 Find the value of each expression.

1) If 2 − x = −2, find the value of 2(2 − x).

2) If x − 1 = −0.4, find the value of 3(x + 0.4).

3) If 8x = −1.6, find the value of 2 + 2x.

4) If 18x = 9, find the value of 6(x − 2).

5) If 5x = 0.5, find the value of 11 − 10x.

6) If −4x = 2, find the value of 2.5(2x − 3).

7) If $2x^2 − 1 = 4$, find the value of $4x^2 − 2$.

8) If $−1 − x^2 = 1.5$, find the value of $2(x^2 + 1.5)$.

Exercises 7 Solve each equation.

1) $1 + \dfrac{x}{2} = 2$

2) $\dfrac{x}{4} + 5 = 6$

3) $4 + \dfrac{x}{8} = 4\dfrac{1}{4}$

4) $\dfrac{x}{5} + 3 = 3\dfrac{3}{5}$

5) $\$34.45 + \dfrac{x}{6} = \$72.95$

6) $\dfrac{x}{2} + \$3.53 = \$16.64$

7) $\dfrac{y}{8} + 4 = -2$

8) $-10 + \dfrac{y}{2} = 15$

Exercises 8 Find the value of each expression.

1) Given $x = -1$, find the value of $2(x - 1)$.

2) Given $x + 1 = 4$, find the value of $x + (-8)$.

3) Given $-2x = 6$, find the value of $3(2 + 2x)$.

4) Given $2(y - 2) = 6.4$, find the value of $3(y - 2)$.

5) If $x + 0.5 = \dfrac{1}{4}$, find the value of $8 - 4(x + \dfrac{1}{2})$.

6) If $2(2 - 3y) = 16$, find the value of $2.5(2 - 3x)$.

7) If $x^2 = 4$, find the value of $12 - 0.2x^2$.

8) If $2y^2 = 18$, find the value of $2(y^2 - 5)$.

Exercises 9 Solve each equation.

1) $y \div 3 + 4 = -2$

2) $-4 + y \div 2 = 5$

3) $-x + 10 = 11$

4) $-8 + y \div 3 = -10$

5) $-3.6 + \dfrac{x}{6} = 2.4$

6) $\dfrac{2x}{3} + 6 = 18$

7) $4.3 + \dfrac{x}{3} = -6.1$

8) $\dfrac{4x}{5} + 1 = 5$

9) $1\dfrac{1}{3}(x) + 2 = -4$

10) $\dfrac{5x}{2} + 21 = -9$

Exercises 10 Find the value of each expression.

1) If $\dfrac{x}{2} = 2$, find the value of $2(1 - x)$.

2) If $\dfrac{2}{x} = 8$, find the value of $4 - 4x$.

3) If $\dfrac{2x}{3} = 0.2$, find the value of $1.8 + 2x$.

4) If $\dfrac{2}{3x} = 1$, find the value of $3(x + 1.6)$.

5) If $\dfrac{2 - x}{5} = -\dfrac{2}{5}$, find the value of $8 - 4x$.

6) If $\dfrac{1}{3x + 1} = 2$, find the value of $6x - 1$.

7) If $\dfrac{x + 3}{2} = 1$, find the value of $2 - 0.2(x + 3)$.

8) If $x = \dfrac{5x - 3}{2}$, find the value of $x + 2$.

Exercises 11 Solve each equation.

1) $8 - \dfrac{2y}{3} = y$

2) $\dfrac{4y}{5} - 9 = y$

3) $\dfrac{x}{4} - 9 = 3x$

4) $8 - \dfrac{x}{8} = -2x$

5) $-2 - \dfrac{y}{2} = 5y$

6) $\dfrac{y}{3} - 1 = -y$

7) $-0.2 - \dfrac{y}{6} = y$

8) $x - 1 = \dfrac{2x}{3}$

Exercises 12 Find the value of each expression.

1) If $x = \dfrac{x-1}{2}$, find the value of $2.5(1-x)$.

2) If $x = \dfrac{4x-3}{3}$, find the value of $2x + 3$.

4) If $x + 1 = \dfrac{x+1}{2}$, find the value of $(2+x)$.

3) If $1 - x = \dfrac{x-4}{2}$, find the value of $3(y+2)$.

5) If $x + \dfrac{1}{5} = \dfrac{x-3}{5}$, find the value of $5 - 2x$.

6) If $1 - \dfrac{x}{2} = \dfrac{x-2}{2}$, find the value of $\dfrac{1}{2}(x-3)$.

7) If $\dfrac{x-1}{2} = \dfrac{2x-3}{2}$, find the value of $1 - x$.

8) If $\dfrac{x+1}{6} = \dfrac{2x-1}{3}$, find the value of $\dfrac{5}{6}(x-4)$.

* Solving Problems

Exercises 13   Solve each problem using the given information.

* Use the following information for Exercise **1-2** below. Matt receives his salary from his company every two weeks. Currently he has $453.71 in his bank account. When he checks his bank account 3 months later, it shows that he now has $3261.54.

1) Which of the following describes how much money he receives from his company every two weeks?

2) What is the value of x?

* Use the following information for Exercise **3-4** below. Oliver has $150.30 in his account. He starts a part-time job during the summer break at a restaurant and will receive $50.45 every week.

3) Which of the following describes how much money he will have after 12 weeks?

4) What is the value of x?

* Use the following information for Exercises **5-6** below. Nick borrowed 12 books from the library. He notices that he borrowed $\frac{1}{3}$ times as many nonfiction books than fiction books.

5) What is the equation used to find the number of books he has left?

6) If he returned 2 nonfiction books to the library, how many books does he have nonfiction books?

* Use the following information for Exercises **7-8** below. Elis spent 4 hours in the garden. She planted some seeds and watered the plants for $1\frac{1}{4}$ hour, and then spent $2\frac{1}{2}$ hour weeding the garden. For the rest of the time, she trimmed the garden.

7) What is the equation used to find the time she spent trimming the garden?

8) For how many hours did she trim the garden?

* Use the following information for Exercises **9-10** below.
  An ice cream store is selling ice cream cones. On Sunday, they sold some vanilla cones, 10 less strawberry cones than vanilla cones, and 11 more chocolate cones than vanilla cones. In total, the store sold 76 cones that day.

9) What is the equation that describes how many other flavors they sold? Use x to represent the number of the other ice cream flavors.

10) What is the value of x?

* Use the following information for Exercises **11-12** below.
  There are 49 pieces of peppermint candy in the bowl. Kelly eats 2 pieces of candy and gave some candy to her friend. By the end of the day, there are 23 pieces left.

11) Which of the following correctly expresses the problem?

12) How many pieces of candy did Kelly give to her friend?

SELF-TEST

1. Which of the following expressions represents "the sum of 2 times x and 2"?

   A. $2 \div x$ **B.** $2(x + 1)$
   C. $2(2 + x)$ **D.** $2 = x$

2. What is the value for x for the equation below?
   $$13 - 2x = 19$$

   A. 3   **B.** 6
   C. -3 **D.** -6

3. What is the value for x for the equation below?
   $$2 + \frac{2}{5}x = 9$$

   A. 20   **B.** 40
   C. -20 **D.** 35

4. What is the value for x for the equation below?
$$\frac{1}{4}(29 + x) = 15$$

A. 1	**B.** 31
C. −1	**D.** 30

5. Which of the following represents "the difference of 3 times x and 7"?

A. $3 \div 7x$	**B.** $3(x - 7)$
C. $3(x + 7)$	**D.** $3x - 7$

6. Which of the following represents "one-half times x squared"?

A. $2x^{\frac{1}{2}}$	**B.** $\frac{1}{2}x^2$
C. $\frac{1}{2}2^x$	**D.** $\frac{1}{2}\sqrt{x}$

7. Given "the sum of 28 and N increased by 5 is 56", what is the value of N?

A. 26	**B.** 23
C. 13	**D.** 41

8. Which of the following represents "2 less than 3 times x"?

A. $2 - 3x$	**B.** $x(3 - 2)$
C. $3(x + 2)$	**D.** $3x - 2$

9. What is the value of x for the equation below?
$$2x - 4 = 18$$

A. 9	**B.** 10
C. 11	**D.** 12

10. What is the value of x for the equation below?
$$2(x - 4) = 18$$

A. 10	**B.** 11
C. 12	**D.** 13

11. What is the value of x for the equation below?
$$\frac{1}{2}(x - 4) = 18$$

      A. 13   **B.** 16
      C. 14   **D.** 40

12. Which of the following represents the expression "the difference of –2 times x and 5 is –12"?

      A. $-2 \div 5x = -12$ **B.** $-2x - 5 = -12$
      C. $-2 + 5x = -12$ **D.** $-2 - 12 = 5x$

13. What is the value of x for the equation below?
$$8x - 8 = -48$$

      A. 5   **B.** 7
      C. –5   **D.** 40

*   Use the following information for Questions **14-15**.
Dave puts different amounts of candy in 3 paper bags. The first bag contains 8 pieces of candy. The second bag has half the amount of candy in the first bag. He does not remember how many pieces he put into the third bag.

14. What is the expression of the equation that represents the problem given that he has 30 total pieces of candy?

      A. $8 + 8 \times 2 + k = 30$ **B.** $30 + k = 8 + 8 \times \frac{1}{2}$

      C. $8 + \frac{1}{2} \times 8 + k = 30$ **D.** $8 + 8 \times 2 - 30 = k$

15. How many pieces of candy does Dave have in total?

      A. 12   **B.** 18
      C. 16   **D.** 14

*   Use the following information for Questions **16-17**.
Julia bought apples, pears, and oranges at a grocery market, which are. She bought 4 times as many apples than pears and $\frac{1}{4}$ as many oranges than pears.

16. If she bought 21 fruits, what is the equation used to determine how many oranges and pears there are?

      A. $21 = 4x + y - \frac{1}{4}z$ **B.** $21 = 4x + x + \frac{1}{4}x$

      C. $0 = 21 - 4x + x + \frac{1}{4}x$ **D.** $21 + 4x + y + \frac{1}{4}z = 0$

17. How many oranges did she buy?

    A. 1   **B.** 2
    C. 3   **D.** 4

18. On Monday, it rained $\frac{1}{4}$ inches. On Tuesday, it rained 2 inches. In total, it rained $8\frac{3}{4}$ inches for five consecutive days. Determine how many inches it rained for the next three days after Tuesday.

    A. 5   **B.** $6\frac{1}{2}$
    C. $5\frac{1}{4}$ **D.** 6

* Use the following information for Questions **19-20**. Will bought 28 bottles of water and drank two bottles on the same day. For the next 5 days, he drank 3 bottles every day.

19. Which of the following equations can be used to determine how many bottles of water are left?

    A. $28 = x + 2 + 3$ **B.** $28 = x + 2 + 5$
    C. $28 = x + 2 + 5(3)$ **D.** $28 + 2 + 5(3) = x$

20. How many bottles of water are left in his refrigerator?

    A. 9   **B.** 10
    C. 11   **D.** 12

* Use the following information for Exercises **21-22**. Cark has 15 chocolate bars. He eats 4 bars and gives his friend 2 more bars than what he just ate.

21. Which of the following shows how many bars are left?

    A. $4 + 4x = 15$ **B.** $4 + x + (2 + 4) = 15$
    C. $4 + (4 - 2) + x = 15$ **D.** $x + 4 + 2 = 15$

22. How many bars does he have left?

    A. 3   **B.** 5
    C. 7   **D.** 9

23. Which of the following represents the equation "the sum of -3 times x and 4 is 9"?

    A. $-3 \div 4x = 9$ **B.** $-3 - 4x = 9$
    C. $-3 + 9 = 4x$ **D.** $-3x + 4 = 9$

24. What is the value of x?

$$-24 + 2x = 15$$

    A. 17   **B.** 19
    C. 21   **D.** 23

25. If $x = 6$, what is the value of $1 - x(y - 2) = -5$?

    A. 1   **B.** 2
    C. 3   **D.** 4

26. If $x = 2$, what is the value of $2xy - \frac{1}{2}x = 7$?

    A. 1   **B.** 2
    C. 3   **D.** 4

27. If $\frac{x}{3} = \frac{2x - 1}{9}$, what is the value of $x - (xy + 2) = 1$?

    A. 1   **B.** 2
    C. 3   **D.** 4

28. If $\frac{1}{2x} = \frac{2}{3x - 1}$, what is the value of $-5x - 7$?

    A. 1   **B.** 2
    C. 3   **D.** 4

29. If $x - 1 = \frac{x - 1}{3}$, what is the value of $-(x - 2)$?

    A. 1   **B.** 2
    C. 3   **D.** 4

30. If $6x^2 = 1$, what is the value of $3x^2 + \frac{1}{2}$?

    A. 1   **B.** 2
    C. 3   **D.** 4

2. Solving Equations with a Variable: Multiplication and Division

3–5. Solve the equation.

$$4 \times x = 7.6$$

SOLUTION

To solve for x, you should divide each side by 4. So, the left side of the equation can
be canceled out. Therefore, only x remains. On the right side, 7.6 is divided by 4.

$4 \times x = 7.6$	Original equation.
$\dfrac{1}{4} \times \overset{1}{4} \times x = 7.6 \times \dfrac{1}{4}$	Divide each side by **4**.
$x = \dfrac{7.6}{4}$ or 1.9	Simplify.

So, the solution of $4 \times x = 7.6$ is 1.9.

Also you can check the solution by substituting x in the equation $4 \times x = 7.6$.

$4 \times 1.9 = 7.6$	Substitute x with 1.9.
$7.6 = 7.6$	This means that the value of x is 1.9.

Quick Exercises 5   Solve each equation.

1) $2 \times n = -8$ 2) $(-2) \times n = 12.8$

3) $15 = y \times 2.5$ 4) $(-4) \times x = -22$

3–6. Solve the equation.

$$10 \div n = 2.5$$

SOLUTION

You can solve the problem using two ways. The answer will be the same when using both
methods.

a)

$10 \div n = 2.5$	Original equation.
$\dfrac{10}{n} = 2.5$	Rewrite as a fraction.
$n \times \dfrac{10}{n} = 2.5 \times n$	Multiply each side by n.
$\overset{1}{n} \times \dfrac{10}{n_1} = 2.5 \times n$	Simplify.

$10 = 2.5n$ Simplify.

$\dfrac{1}{2.5} \times 10 = 2.5n \times \dfrac{1}{2.5}$ Multiply each side by $\dfrac{1}{2.5}$.

$\dfrac{1}{2\cancel{.5}_1} \times \overset{4}{\cancel{10}} = \overset{1}{\cancel{2.5}}n \times \dfrac{1}{2\cancel{.5}_1}$ Simplify with the GCF*.   * The GCF of 2.5 and 10 is 4.
                                                                            The GCF of 2,5 and 2.5 is 1.

$4 = n$

So, the value of n is 4.

b) Given an equation like $10 \div n = 2.5$, you can rewrite it so that $10 = 2.5n$.

$10 \div n = 2.5$ Original equation

$10 = 2.5n$ Rewrite the equation as $10 \div n = 2.5$.

$\dfrac{1}{2.5} \times 10 = 2.5n \times \dfrac{1}{2.5}$ Multiply each side by $\dfrac{1}{2.5}$.

$\dfrac{1}{2\cancel{.5}_1} \times \overset{4}{\cancel{10}} = \overset{1}{\cancel{2.5}}n \times \dfrac{1}{2\cancel{.5}_1}$ Simplify with the GCF*.   * The GCF of 2.5 and 10 is 4.
                                                                            The GCF of 2,5 and 2.5 is 1.

$4 = n$

So, the value of n is 4.

You can solve the problem both ways and the answer will be identical.

c) Once you solved the equation, you can check your solution by substituting in the value in the original equation.

$10 \div n = 2.5$ Original equation

$10 \div 4 = 2.5$ Substitute n with 4.

$2.5 = 2.5$

The value of n is 4.

* Remember that you can manipulate the equations.

dividend ÷ divisor = quotient

dividend = divisor × quotient

Quick Exercises 6   Solve each equation.

1) $(-5) \div d = 5$ 2) $c \div 7 = (-3.5)$

3) $x \div 1.5 = 12$ 4) $5 = (-5.5) \div x$

3–7. Solve the equation.

$$6 \div \frac{x}{2} = 4$$

SOLUTION

You can solve the problem using two ways. The answer will be the same for both methods.

a)

$6 \div \left(\frac{x}{2}\right) = 4$ i) Flip the numerator and denominator (reciprocals). $6 \times \frac{2}{x} = 4$

ii) Change the operation.
iii) Then multiply.

$\dfrac{12}{x} = 4$ Find the reciprocal of the divisor and multiply.

$x \times \dfrac{12}{x} = 4 \times x$ Multiply each side by x.

$12 = 4x$ Simplify. $(x \times \frac{1}{x}) = 1$

$\dfrac{1}{4} \times 12 = 4x \times \dfrac{1}{4}$ Multiply each side by $\frac{1}{4}$.

$3 = x$ Simplify with the GCF. $(4x \times \frac{1}{4}) = x$

b) When solving an equation like $6 \div \frac{x}{2} = 4$, you can rewrite it to be $6 = \frac{x}{2} \times 4$.

$6 \div \dfrac{x}{2} = 4$ or $6 = \dfrac{x}{2} \times 4$ Rewrite the equation.

$6 = 2x$ Simplify with the GCF. $(\frac{x}{2} \times 4) = 2x$

$\dfrac{1}{2} \times 6 = 2x \times \dfrac{1}{2}$ Multiply each side by ½.

$3 = x$ Simplify with the GCF. $\frac{1}{2} \times 6 = 3$, $(2x \times \frac{1}{2}) = x$

So, the solution of $6 \div \dfrac{x}{2} = 4$ is $x = 3$.

* Reciprocals: Two numbers are reciprocals if their product is 1. For example;

$$\frac{1}{2} \times \frac{2}{1} = 1 \qquad \frac{5}{7} \times \frac{7}{5} = 1 \qquad \frac{9}{10} \times \frac{10}{9} = 1$$

Quick Exercises 7   Solve each equation.

1) $y \div \dfrac{1}{2} = 5$

2) $4 \div \dfrac{x}{2} = 6$

3) $\dfrac{2}{3}x \div 3 = 5$

4) $\dfrac{1}{2}(4 \div y) = \dfrac{1}{6}$

3-8. Solve the equation.

$$\frac{2}{3}(x) = -4 \text{ or } 2x \div 3 = -4$$

SOLUTION

$\frac{2}{3}(x) = -4$	Original equation.
$(\frac{3}{2} \times \frac{2}{3})(x) = -4 \times \frac{3}{2}$	Multiply $\frac{3}{2}$ on both sides.
$x = -\frac{12}{2} \text{ or } -6$	Simplify. $(\frac{3}{2} \times \frac{2}{3}) = 1$

So, the solution of $\frac{2}{3}(x) = -4$ is $-\frac{12}{2}$ or -6.

You can check the solution of $x = -6$ by substituting the value in the original equation.

$\frac{2}{3}(x) = -4$	Original equation.
$\frac{2}{3}(-6) = -4$	Substitute x with -6.
$-\frac{12}{3} = -4 \text{ or } -4 = -4$	This means the value is correct.

Quick Exercises 8   Solve each equation.

1) $3y = \frac{1}{10}$

2) $\frac{1}{8}x = -2$

3-9. Solve the equation.

If $\frac{4}{x} = 2$, what is the value of $4x \div 2$?

SOLUTION

First, find the value of x and then find the value of the expression.

$\frac{4}{x} = 2$	Given equation
$(x \times \frac{4}{x}) = 2 \times x$	Multiply each side by x.
$4 = 2x$	Simplify. $x \times \frac{4}{x} = 4$
$2 = x$	Divide each side by 2 and simplify.

Now, find the value of $4x \div 2$.

$(4 \cdot 2) \div 2$	Substitute x with 2.

So, the value of $4x \div 2$ is 4.

Exercises 14 Solve each equation.

1) $4 \times n = -56$

2) $-1 \times n = 9$

3) $y \times (-7) = 42$

4) $-2.6 \times x = -18.2$

5) $z \times (-3.5) = 14$

6) $-1.4n = 58.8$

7) $2 + y \times (-2) = -3$

8) $-2 + (-2 \times x) = -7$

9) $z \times (-5) + 3 = 18$

10) $-3n - 2 = 22$

Exercises 15 Solve each equation.

1) $(2 \times n) - 1 = 7$

2) $(1 \times n) + 9 = -1$

3) $2 - (y \times 3) = 11$

4) $1 + (2 \times x) = -5$

5) $2(z \times 5) - 7 = 23$

6) $9(n + 1) = -18$

7) $6a - 5 = 43$

8) $2c + 3 = 15$

9) $(-2)^2 y = 36$

10) $x(-1)^3 = 49$

Exercises 16 Find the value of each expression.

1) Given $x = 2.4$, find the value of $2(8 \times x)$.

2) Given $c = 2$, find the value of $14 \div c + 8$.

3) Given $x = 0.3$, find the value of $9 \div x$.

4) Given $y = 1.4$, find the value of $3(y \times 1.5)$.

5) If $n = 0.5$, find the value of $8 \times 4n$.

6) If $c = 1.2$, find the value of $2.5(c \div 3)$.

7) If $x = 4$, find the value of $(12 \div 0.2)x$.

8) If $y = 5.5$, find the value of $2(y \times 2.0)$.

9) If $x = \dfrac{1}{2}$, find the value of $6 \div (4x)$.

10) If $3y = 2$, find the value of $(y \times 9) + 2$.

Exercises 17 Solve each equation.

1) $x \div 3 = 12$

2) $-4.8 \div n = 2$

3) $0.5(4 \div d) = 8.0$

4) $0.2(c \div 8) = 5$

5) $5x \div 5 = 15$

6) $2(1 \div a) = -7$

7) $(y - 2) \div 2 = 5$

8) $4 \div (0.5x) = 8$

9) $(x + 2) \div 3 = -5$

10) $0.5(28 \div y) = 7$

Exercises 18 Find the value of each expression.

1) Given $x = -1$, find the value of $2(2 \times x^2)$.

2) Given $x = -2$, find the value of $4 \div x - 3$.

3) Given $x - 0.6 = 12$, find the value of $\frac{1}{3}(x - 0.6)$.

4) Given $(y - 2)^2 = 4$, find the value of $3(y - 1)$.

5) If $3x - \sqrt{5} = 5$, find the value of $\frac{1}{5}(3x - \sqrt{5})$.

6) If $x - \sqrt{4} = 2$, find the value of $2(x - \sqrt{4})^2$.

7) If $x = \sqrt{4}$, find the value of $(64 \div 4)x$.

8) If $(y - 1) = 1$, find the value of $3 - (y \times \frac{1}{2})^2$.

Exercises 19 Solve each equation.

1) $(8 \div n) - 1 = 7$

2) $(1 \div n) + 1 = -1$

3) $5(z \div 5) - 2 = 3$

4) $9 \div (n + 1) = -3$

5) $\frac{4}{5}a - 1 = 3$

6) $\frac{2}{3}c + 3 = 5$

7) $(\frac{1}{6}y) = 1\frac{2}{3}$

8) $\frac{4}{5}(x - 1) = 1\frac{6}{10}$

9) $1\frac{1}{2x} = 1\frac{1}{8}$

10) $(x - 1) = (\frac{1}{2})^2$

Exercises 20 Find the value of each expression.

1) Given $x = 2.4$, find the value of $2(2 \times x)$.

2) Given $c = 2$, find the value of $14 \div c - 8$.

3) Given $x = -0.3$, find the value of $9 \div x$.

4) Given $-y = 1.4$, find the value of $3(y \times 1.5)$.

5) If $2x + 1.5 = -0.5$, find the value of $-8 \times 4x$.

6) If $-(x \div 2) = 0.2$, find the value of $2.5(x \div 2)$.

7) If $(2x)^2 = 4^2$, find the value of $12 \div x$.

8) If $(y - 5)^2 = (-1)^2$, find the value of $(y - 5) \times 2.0$.

9) If $-x = \dfrac{1}{2}$, find the value of $-6 \div (4x)$.

10) If $(3y) = -(3)^2$, find the value of $y \times 3 + 11$.

Exercises 21 Solve each equation.

1) $2x - 3 = -1$

2) $-y + 7 = -5$

3) $4 - 2y = 16$

4) $2 - 7x = -5$

5) $\dfrac{x + 1}{2} = 2(x - 1)$

Hint: First, multiply 2 on both sides.

6) $\dfrac{3(y + 2)}{2} - 1 = 2y + 4$

Hint: Add 1 to both sides.

7) $\dfrac{5}{2x + 1} = -5$

Hint: i) $\dfrac{a}{b} = c$, ii) $\dfrac{a}{c} = b$, or iii) $a = bc$

8) $\dfrac{1}{2(y - 1)} + 2 = 1$

Exercises 22 Solve each equation.

1) $24 \div \dfrac{y}{2} = 4$

2) $18 \div \dfrac{y}{7} = 21$

3) $\dfrac{2}{3} \div \dfrac{x}{8} = -8$

4) $\dfrac{1}{8} \div \dfrac{y}{2} = -4$

5) $\dfrac{y}{2} \div 5 = \dfrac{1}{5}$

6) $4 \div \dfrac{2}{x} = \dfrac{4}{8}$

7) $5 \div \dfrac{4}{x} = 1\dfrac{1}{4}$

8) $20 \div \dfrac{1}{x} = \dfrac{1}{2}$

Exercises 23 Find the value of each expression.

1) Given $x = -1$, find the value of $3(4 \div x^2)$.

2) Given $4x = 2$, find the value of $4 \div x - 1$.

3) If $\dfrac{2x}{3} = \dfrac{1}{2}$, find the value of $\dfrac{1}{3}(36 \div x)$.

4) If $(y \div 2) = 4$, find the value of $3(y \div 2)$.

5) If $2(x \div \sqrt{2}) = 3$, find the value of $\dfrac{2}{3}(\dfrac{x}{\sqrt{2}})$.

6) If $2(x \div \sqrt{25}) = 4$, find the value of $2(x \div \sqrt{25})^2$.

7) If $x = \sqrt{4}$, find the value of $16 \div (2x)$.

8) If $2y(4 \div 2)^2 = 8$, find the value of $3 - (2 \div y)$.

* Solving Problems

Exercises 24   Solve each problem using the given information.

* Use the following information for Exercises **1-3** below. A miser is organizing his coins. He has quarters and dimes in several bags. Each bag has 7 less quarters than dimes.

1) What equation could be used to solve how much money is in each bag? Use x to represent the number of quarters and y to represent the number of dimes. Show your work.

2) What are the values of x and y if there are 12 bags of coins?

3) How much money is there in 12 bags?

* Use the following information for Exercises **4-5** below. At a fundraiser, 9 people donated some money. Eight of them donated the same amount of money, except for one person who donated $25.00 more than the others.

4) If the money they donated adds up to $325.00, what is the equation showing how much each person donated? Use x to represent the amount of money.

5) How much money did each person donate?

* Use the following information for Exercises **6-7** below. Bob has 105 toy cars that he would like to keep in several boxes. He puts 8 toy cars in each box except for one, which he puts in half as many cars than the other boxes.

6) Write the equation than can be used to find how many boxes he uses to keep his cars.

7) How many boxes does Bob use?

8) The school is planning a field trip. There are 410 students in total and 35 students in each bus, but two buses have 5 less students than the other buses. Write the equation used to determine many buses will be needed to take every student.

9) How many buses will be needed to take every student?

1. Which of the following expressions represents "the product of $\frac{1}{3}$ and x"?

    A. $\frac{1}{3} \div x$ **B.** $\frac{1}{3} - x$

    C. $\frac{1}{3} + x$ **D.** $\frac{1}{3} \times x$

2. Which of the following expressions represents "the product of $\frac{1}{2}N$ and 1"?

    A. $\frac{1}{2}N + 1$ **B.** $N \div 1$

    C. $\frac{1}{2}N$ **D.** $N - 1$

3. What is the value of x for the equation below?

$$\frac{1}{5}(9 \times x) = -3$$

    A. $1\frac{2}{3}$ **B.** $-1\frac{2}{3}$

    C. $\frac{3}{5}$ **D.** $-\frac{3}{5}$

4. What is the value of x for the equation below?

$$-9 \times x = 3\frac{3}{5}$$

    A. $-\frac{2}{5}$ **B.** $\frac{2}{5}$

    C. $-2\frac{1}{2}$ **D.** $2\frac{1}{2}$

5. What is the value of x for the equation below?

$$-3 \times \frac{1}{5}x = -9$$

    A. 15   **B.** -15

    C. $-1\frac{2}{3}$ **D.** $1\frac{2}{3}$

6. Which of the following expressions represents "$\frac{1}{5}$ divided by x"?

    A. $\frac{1}{5} \div x$ **B.** $\frac{1}{5} - x$

    C. $\frac{1}{5} + x$ **D.** $\frac{1}{5} \times x$

7. Which of the following expressions represents "$\frac{1}{2}y$ divided by 3"?

    A. $3 \div y$ **B.** $\frac{1}{2}y \div 3$

    C. $\frac{1}{2} + y$ **D.** $\frac{1}{2}y \times 3$

8. Which of the following represents the expression "the quotient of 2 times x and 2 is 6"?

    A. $2x \div 2 = 6$ **B.** $2 - 2x = 6$
    C. $2 + 2x = 6$ **D.** $2 \times 2x = 6$

9. Which of the following represents the expression "the quotient of 81 and 3 is y"?

    A. $3 \div 81 = y$ **B.** $y = 81 \div 3$
    C. $3 + y = 81$ **D.** $y = 81 \times 3$

10. What is the value of x for the equation below?
$$\frac{1}{3}(9 \div x) = 1$$

    A. 1   **B.** 2
    C. 3   **D.** 4

11. What is the value of x for the equation below?
$$\frac{1}{2}x \div 4 = 6$$

    A. 6   **B.** 12
    C. 32   **D.** 19

12. What is the value of x for the equation below?
$$(9 \div x) = 1\frac{1}{3}$$

A. $4\frac{1}{2}$ **B.** $\frac{2}{9}$

C. 8   **D.** $3\frac{2}{3}$

13. What is the value of x for the equation below?
$$\frac{1}{2}x \div 2 = 2$$

A. 4   **B.** 8
C. 12   **D.** 24

14. What is the value of "the product of 4 times x and 2 is 54"?

A. 9   **B.** 13.5
C. 27   **D.** 6.75

15. What is the value of "the product of N and 1 is 14"?

A. −14   **B.** 14
C. −7   **D.** 7

16. What is the value for x for the equation below?
$$\frac{x}{4} = -20$$

A. 5   **B.** −5
C. 80   **D.** −80

17. David and his four friends went a restaurant and equally shared the bill of $79.85 for their meal. Write an expression describing how much money each person gave. Use x to represent the amount of money.

A. $5 \div \$79.85 = y$ **B.** $y = \$79.85 \div 5$
C. $5 + y = \$79.85$ **D.** $y = \$79.85 \times 5$

18. What is the value of x for the equation below?
$$5 \div (7x) = \frac{5}{7}$$

A. −5

B. 7

C. $1\frac{24}{25}$

D. 1

19. What is the value of x for the equation below?

$$x \div 8 = 4\frac{3}{8}$$

A. 35

B. $1\frac{29}{35}$

C. $\frac{35}{64}$

D. 8

* Use the following information for Questions **20-21**. Joey has 21 toy cars he needs to put in several boxes. He puts 6 cars in each box and has 3 cars remaining.

20. Which of the following equations could be used to find how many boxes he used?

A. $(21 \div x) + 3 = 6$

B. $21 \times 6 = x + 3$

C. $(21 - 3) \times x = 6$

D. $(21 - 3) \div x = 6$

21. How many boxes did he use?

A. 2

B. 4

C. 5

D. 7

22. If a circumference of a circle is $1\frac{1}{4}\pi$ cm, write an expression that describes the radius of the circle. Use r to represent the radius.

A. $r\pi = 1\frac{1}{4}$

B. $2r = 1\frac{1}{4}$

C. $r = 1\frac{1}{4}$

D. $2r = 1\frac{1}{4}\pi$

23. Marcus is dividing his marbles into 2 jars, which are marked JA and JB. JB has 5 more marbles than JA. If he has 18 marbles, which expression could be used to show how many marbles are in each jar?

A. $5 = 21 \times x$

B. $y = 6x + 21$

C. $21 = 5x$

D. $21 = (5 + x) + x$

24. At a grocery market, one bunch of bananas cost $3.20. If Judy paid $12.80 for bananas, which equation could be used to find the number of bunches she bought?

A. $12.80 = x + 3.20$
B. $12.80 = 3.20x$
C. $12.80 = x \div 3.20$
D. $3.20 = 12.80 \times x$

25. Jack is counting his quarters. He has organized his coins so that there are 6 quarters in a bag. If Jack has $9 worth of quarters in total, what equation could be used to find how many bags he has?

A. $9 = 6(0.25)x$
B. $9 = x \div 6$
C. $9 \times 6(0.25) = x$
D. $9 = 0.25 \div x$

26. At a school, 79 students and 7 teachers are going a field trip. A bus can seat 28 people. Which equation could be used to show how many buses are needed to take every student and teacher?

A. $(79 + 7) \div x = 28r2$
B. $79 + 7 = 28x$
C. $(79 + 7) \div x = 28 + \dfrac{2}{x}$
D. All of the above.

27. Use the information from Question 26. How many buses will be needed to take every student and teacher?

A. 2
B. 3
C. 4
D. 5

28. If the quotient of an equation is 6, what is the value of the divisor given a dividend of 24?

A. 3
B. 4
C. 6
D. $\dfrac{1}{4}$

29. Two tables are set out for a dinner. On each table are two plates of samosas. The first table has $\dfrac{1}{3}$ as many samosas than the second table. If there are 12 samosas in total, what is the equation used to find how many samosas there are on the second table?

A. $\dfrac{1}{3}x - x = 12$
B. $\dfrac{1}{3}x = 12$
C. $\dfrac{1}{3}x + x = 12$
D. $(\dfrac{1}{3} + x) + x = 12$

30. Marcus is dividing 54 pieces of M&Ms candy between his two brothers. His younger brother will get half as much as his older brother. Which expression could be used to show how many pieces each brother will get? Use x to represent the number of M&Ms.

A. $(\frac{1}{2} + 1)x = 54$

B. $x = \frac{1}{2}x + 54$

C. $54 = \frac{1}{2}x$

D. $54 = 2 \times x + 1$

31. The bakery is baking two kinds of loaves of banana bread. The first loaf has 6 bananas and the second loaf has twice as many bananas as the first loaf. What is the equation showing how many bananas are in the first and the second loaves? Use x to represent the number of bananas in the first loaf and y to represent the number of bananas in the second loaf.

A. $6x + \frac{1}{2}y$

B. $6(x + 2y)$

C. $\frac{1}{2}x + 6y$

D. $2(6x) + 6y$

* Use the following information for Questions **32-34.** A miser is organizing his coins. He is putting quarters and dimes in several bags. Each bag has 5 more quarters than dimes.

32. What equation could be used to solve how much money is in each bag? Use x to represent the number of quarters and y to represent the number of dimes.

A. $5 + 0.25x + 0.1y$

B. $(5)(0.25) + 0.25x) + 0.1y$

C. $0.25x + (5 + 0.1y)$

D. $5(0.25x) + 0.1y$

33. What are the values of x and y if there are 8 bags?

A. $x = 3.25, y = 0.80$

B. $x = 2.25, y = 0.80$

C. $x = 4.25, y = 0.80$

D. $x = 3.25, y = 1.25$

34. How much money is there in 8 bags?

A. $4.00

B. $4.05

C. $5.05

D. $4.25

* Use the following information for Questions **35-37.** At a fundraiser, 9 people donated some money. Eight of them donated the same amount of money except for one person, who donated $12.00 less than the others.

35. If the money they donated adds up to $250.00, how did you set up the equation to show how much money each person donated? Use x to represent the money that was donated.

A. $250 = (x + 12) + 9x$

B. $250 = (x - 12) \div 9x$

C. $250 = x + 9x$

D. $250 = (x - 12) + 9x$

36. How much money did each person donated, not including the one who donated less?

      A. $25.40   **B.** $26.20
      C. $27.80   **D.** $15.70

37. A table is set out for a buffet. On the table are plates of crab. Each plate has exactly 5 pieces of crab except for a plate that has twice as many pieces. There are 224 pieces of crab in total. What is the equation used to find how many plates of crab are set out?

      A. $5x(1 + 2) = 224$ **B.** $224 = x \div 5$
      C. $5x = 224$ **D.** $224 = 5x + 2$

* Use the following information for Questions **38-39**. Sam has 92 toy cars that he would like to keep in several boxes. He wants to put in 17 toy cars in each box except for one, which he wants to have 2 more cars than the other boxes.

38. Which of the following equations could be used to find how many boxes he needs to accommodate his toys?

      A. $92 = 17x + (2 - x)$ **B.** $92 = 17x + (x + 2)$
      C. $92 = 17x + (x - 2)$ **D.** $92 = 17x + x$

39. How many boxes does he use?

      A. 3   **B.** 4
      C. 5   **D.** 6

40. The school is planning a field trip. There are 340 students in total and 35 students in each school bus, but two school buses each have 5 less students than the other buses. Which equation could be used to show how many buses are needed to take every student?

      A. $340 = [x + 2(x + 5)]35$ **B.** $340 = 35x + 2(5)$
      C. $340 = 35x - 2(5)$ **D.** $340 = [x - 2(5)]35$

41. How many buses will be needed to take every student?

      A. 10   **B.** 9
      C. 11   **D.** 12

42. If the quotient of an equation is 12, what is the value of the divisor given that the dividend is 8?

      A. 6   **B.** 7
      C. 8   **D.** 9

43. If $x = 6$, what is the value of $1 - x(y \div 2) = -5$?

        A. 1   **B.** 2

        C. 3   **D.** 4

44. If $x = 2$, what is the value of $2x - (y \div 2) = 2$?

        A. -1 **B.** -2

        C. -3 **D.** -4

45. If $\dfrac{1}{3} = 2x - 1$, what is the value of $3x + 2$?

        A. 1   **B.** 2

        C. 3   **D.** 4

46. If $\dfrac{1}{x + 1} = \dfrac{2}{3x}$, what is the value of $x + 1$?

        A. 1   **B.** 2

        C. 3   **D.** 4

47. If $x + 1 = \dfrac{4}{x + 1}$, what is the value of $(x + 1)$?

        A. 1   **B.** 2

        C. 3   **D.** 4

48. If $2(x - 3) = 4$, what is the value of $\dfrac{1}{2}(2x - 2)^2 - 1$?

        A. 1   **B.** 2

        C. 3   **D.** 4

3. Solving Equations with Two Variables

3–10. Solve the equation.

$$-2x + 1 = -x$$

> **SOLUTION**
>
> $$\underbrace{-2x + 1}_{\text{Left}} = \underbrace{-x}_{\text{Right}}$$
>
> If both sides of an equation have variables, then you can rearrange it so that one side has all the variables. From there the value of the variable can be found.
>
> | | |
> |---|---|
> | $-2x + 1 = -x$ | Original equation |
> | $\mathbf{2x} - 2x + 1 = -x + \mathbf{2x}$ | Add $2x$ to both sides. |
> | $1 = x$ | Simplify. |
>
> So the value of x is 1.

Quick Exercises 9   Solve each equation.

1) $3x = -3(2x - 1)$

2) $-y = \dfrac{(-4)y - 5}{5}$

3) $\dfrac{1}{3}(y + 1) = -2(2y + 1)$

4) $3(3 - 2x) = -7(2x + 1)$

3–11. If $3x = -7$, what is the value of y in $3x + y = -3$?

> **SOLUTION**
>
> a) There are two variables in the equation, but x is given. So, you can replace $3x$ with -7 and then solve for y.
>
> | | |
> |---|---|
> | $3x + y = -3$ | Original Equation |
> | $-7 + y = -3$ | Substitute $3x$ with $-\mathbf{7}$. |
>
> b) If you would like to solve for y in the equation $-7 + y = -3$, then add 7 to both sides.
>
> | | |
> |---|---|
> | $-7 + 7 + y = -3 + 7$ | Subtract **7** from both sides. |
> | $y = 4$ | Simplify. |
>
> So, the value of y is 4.
>
> You can check the solution by substituting the two variables in the equation $3x + y = -3$.
>
> | | |
> |---|---|
> | $3x + y = -3$ | Original equation |
> | $(-7) + 4 = -3$ | Substitute $3x = -7$, $y = 4$. |
> | $-3 = -3$ | Simplify. |
>
> This means that the value of y is 4.

Quick Exercises 10   Solve each equation.

1) If $n = 5$, $102 = 4n + 2m$

2) If $x = 6$, $81 = 5y + x$

3) If $y = 4$, $2x + 3y = 28$

4) If $x = 4$, $2(2x + y) = 104$

3–12. Solve the equation.

$$\text{If } x = 5, \; 2(x - 3y) = -20$$

SOLUTION

a) First, replace x with 5 and then solve for y.

$2(x - 3y) = -20$	Original equation
$2(5 - 3y) = -20$	Replace x with 5.
$10 - 6y = -20$	Distributive Property

b) Solve the equation for y.

$-6y = -30$	Subtract 10 from each side.
$y = 5$	Divide each side by –6.

So the value of y is 5.

Check to verify the value of y is 5.

$2(x - 3y) = -20$	Original equation
$2[5 - 3(5)] = -20$	Substitute $x = 5$, $y = 5$.
$-20 = -20$	Simplify.

This means that the value of y is 5.

Quick Exercises 11   Solve each equation using the given variables.

1) If $-n = -1$, $4(n - 2m) = -6$

2) If $n = -2$, $-(n - 2m) = -6$

3) If $-n = 2$, $(2n - m) = -6$

4) If $-n = -1$, $4(2n - m) = -6$

Exercises 25 Find the value of the variable in each equation.

1) $-3x + x = -4$

2) $\dfrac{1}{4} + x = -2 - x$

3) $2(x + 2) = -4x$

4) $\dfrac{x}{8} + 5 = \dfrac{x}{4} - 1$

5) $20 - x = -5x$

6) $\dfrac{2}{3} + y = -\dfrac{4}{9} - y$

7) $\dfrac{x}{2} = -\dfrac{1}{2} + 2x$

8) $3x + 1 = -x - 7$

9) $\dfrac{1}{2}x = -10 + x$

10) $2 + 3x = \dfrac{x}{3}$

Exercises 26 Find the value of each expression.

1) Given $x = -2$ and $y = 2$, find the value of $1 + 2xy$.

2) Given $x = -1$ and $y = 3$, find the value of $17 + 2xy + 3x$.

3) If $4x = -16$ and $2y + 1 = 5$, find the value of $xy + 2y$.

4) If $\dfrac{1}{3} = -\dfrac{x}{3}$ and $y = 2$, find the value of $2x + 3y$.

5) If $2x - 3 = -2$ and $2(y - 1) = 2$, find the value of $(2x - 3) + 4(y - 1)$.

6) If $x - 1 = -5$ and $(y - 1) = 2$, find the value of $(x - 1)^2 + 2(y - 1)^2$.

Exercises 27 Find the value of each equation.

1) $-x = -5(x + 2)$

2) $-2y = -\dfrac{4y - 3}{3}$

3) $4x - 3 = -2x$

4) $-3x = -7(2x + 1)$

5) $\dfrac{2x + 5}{2} = 4x$

6) $6x + 1 = -3(x - 2)$

7) $4(x - 3) = -2(x + 12)$

8) $\dfrac{1}{3}(x - 1) = -2(x - 2)$

9) $2(x - 3) + 1 = -2(x + 1)$

10) $-3(x + 2) - 7 = -5(2x - 3)$

11) $\dfrac{3}{5}(x - 1) = -(x - 2) + 2$

12) $6x + 1 = 3(x + 2)$

Exercises 28 Find the value of each equation.

1) If $y = 3$, $x + 2y = 42$

2) If $x = 4$, $2x + y = 114$

3) If $-y = 2$, $128 = 24y + x$

4) If $a = -8$, $3a + 56 = b$

5) If $10 \div n = 5$, $102 = 4n + 2m$

6) If $4 \div x = -2$, $78 = 5y + x$

7) If $y \div 4 = -1$, $2x + 3y = 28$

8) If $x = 4$, $2(2x + y) = 106$

9) If $2y = -1$, $57 = 3(2y + x)$

10) If $a \div 3 = -1$, $3(a + 5) = b$

Exercises 29   Find the value of each expression.

1) If $x = 3$ and $y = -2$, find the value of $x - 4y$.

2) If $x = 8$ and $y = -2$, find the value of $5x - 25 - y$.

3) If $2(b + 4) = 4$ and $3a = 1$, find the value of $4b - 9a + 13$.

4) If $2x = 3$ and $3y = 2$, find the value of $x + 3y$

5) If $\dfrac{x - 4y}{3} = 1$, find the value of $3(x - 4y) - 2$.

6) If $\dfrac{2m - 6n}{2} = -2$, find the value of $2 - 2(m - 3n)$

7) If $\dfrac{y}{3} = 1$ and $\dfrac{x}{2} = -1$, find the value of $2(7y - x)$.

8) If $\dfrac{2}{y} = 3$ and $\dfrac{5}{x} = 2$, find the value of $3(2x - 3) - 3y$.

9) If $\dfrac{1}{2b + a} = 2$, find the value of $3(4b + 2a)$.

10) If $\dfrac{4}{2x - 3y} = 3$, find the value of $6(2x - 3y)$.

Exercises 30   Find the value of each equation.

1) If $y = 2$, $\dfrac{x}{2} + y = 4$

2) If $x = 3$, $2x + \dfrac{y}{3} = 1$

3) If $y = 2$, $3 = \dfrac{y}{4} + x$

4) If $x = 9$, $y + 5 = \dfrac{x}{3}$

5) If $x = 5$, $10 = \dfrac{x}{5} + \dfrac{y}{2}$

6) If $y = 4$, $3 = 5x + \dfrac{y}{2}$

1. If $k = 6$, what is the value of N in the equation $19 - (k + 7) = N$?

    A. 6   **B.** 7
    C. 8   **D.** 9

2. If $c = 6$, what is the value of N in the equation $N - 3c = 7$?

    A. 23   **B.** 24
    C. 25   **D.** 26

3. If $z = 15$, what is the value of k in the equation $19 + k + 2z = 83$?

    A. 24   **B.** 35
    C. 34   **D.** 32

4. If $k = 2$, what is the value of N in the equation $57 - (14k + 7) = N$?

    A. 20   **B.** 22
    C. 24   **D.** 26

5. Add the product of 2 and K to the product of 4 and N. Given that $K = 4$, and the value of the equation is 20, what is the value of N?

    A. 1   **B.** 2
    C. 3   **D.** 4

6. Multiply the sum of 5 and N to the difference of 10 and K. Given that $K = 8$, and the value of the equation is 14, what is the value of N?

    A. 1   **B.** 2
    C. 3   **D.** 4

7. Subtract the quotient of 36 and K to the product of 4 and N. Given that $K = 2$, and the value of the equation is 6, what is the value of N?

    A. 1   **B.** 2
    C. 3   **D.** 4

8. If $\dfrac{2x}{3} = -1$ and $\dfrac{y}{4} = 1$, what is the value of $1 - x(y - 2)$?

    A. 1   **B.** 2
    C. 3   **D.** 4

9. If $2x = 1$ and $y = 2$, what is the value of $2xy - \dfrac{1}{2}y + 2$?

    A. 1   **B.** 2
    C. 3   **D.** 4

10. If $\dfrac{x}{2} = 1$ and $\dfrac{2y + 1}{7} = 1$, what is the value of $(xy - 2) - x$?

    A. 1   **B.** 2
    C. 3   **D.** 4

11. If $\dfrac{4}{y} = 3$ and $\dfrac{3}{x} = 3$, what is the value of $3(2x - 3) + 3y$?

    A. 1   **B.** 2
    C. 3   **D.** 4

12. If $x = -1$ and $2 = \dfrac{4}{y - 1}$, what is the value of $x(y - 5)$?

    A. 1   **B.** 2
    C. 3   **D.** 4

13. If $6 \div x = 2$, what is the value of $3\dfrac{1}{2} + \dfrac{x}{6}$?

    A. 1   **B.** 2
    C. 3   **D.** 4

14. If $x \div 2 = 3$ and $y - 2 = \dfrac{1}{3}$, what is the value of $x(y - 2)$?

    A. 1   **B.** 2
    C. 3   **D.** 4

15. If $6x = 1$ and $\dfrac{y}{6} = \dfrac{1}{2}$, what is the value of $12x + \dfrac{1}{3}y$?

    A. 1   **B.** 2
    C. 3   **D.** 4

4. Converting Measurement Units

3–13. Convert the measurements.

$$21.6 \text{ lb} = \underline{\hspace{3cm}} \text{ ounces}$$

SOLUTION

First, look at the given units of the problem and then set up a proportion of two ratios with the same unit. You should know that 1 pound (lb) = 16 ounces (oz).
i) Set up a ratio using the given information. Let x be the unknown quantity as 21.6 lb = x ounces.

ii) Set up a proportion of two ratios with the same unit and then find the value of x using cross product property.

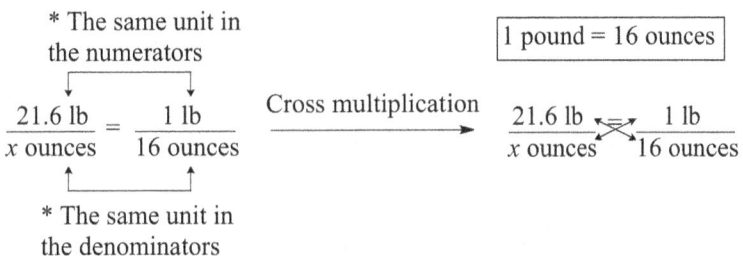

iii) Write a proportion, (21.6 lb)(16 ounces) = (x ounces)(1 lb)

The units can cancel each other out.

$$(21.6 \text{ lb})(16 \text{ ounces}) = (x \text{ ounces})(1 \text{ lb}) \implies (21.6)(16) = (x)(1)$$

The units can cancel each other out.

$$345.6 = x \qquad \text{Multiply 21.6 by 16.}$$
So, 21.6 lb = $\underline{\quad 345.6 \quad}$ ounces

3–14. Convert the measurements.

$$28 \text{ gallons} = \underline{\hspace{3cm}} \text{ quarts}$$

SOLUTION

You should know 1 gallon = 4 quarts.

$$\frac{28 \text{ gallons}}{x \text{ quarts}} = \frac{1 \text{ gallon}}{4 \text{ quarts}} \qquad \text{Set up a proportion of two ratios.}$$

$(28 \text{ gallons})(4 \text{ quarts}) = (1 \text{ gallons})(x \text{ quarts})$ Cross Product Property
$28 \times 4 = x$ Cancel the units.
$112 = x$ Multiply 28 by 4.

So, 28 gallons = $\underline{\quad 112 \quad}$ quarts.

3–15. Conversions

Length	**Volume**
1 yard (yd.) = 3 feet (ft) = 36 inches (in)	1 kiloliter (kL) = 1,000 liters (L)
1 foot = 12 inches	1 hectoliter (hL) = 100 liters (L)
1 meter (m) = 100 centimeters (cm) = 1,000	1 dekaliter (daL) = 10 liters (L)
millimeters (mm) = 1,000,000 micrometers (μm)	1 liter (L) = 10 deciliters (dL)
1 mile (mi.) = 1,760 yards (yd.) = 5,280 feet (ft)	1 liter (L) = 100 centiliters (cL)
1 centimeter = 0.3937 inch (in)	1 liter (L) = 1,000 milliliters (mL)
1 inch = 2.54 centimeters (cm)	
1 foot = 0.305 meter (m)	
1 meter = 3.28 feet (ft)	
1 yard = 0.914 meter (m)	
1 mile = 1.609 kilometers (km)	
Volume	**Mass (weight)**
1 pint = 2 cups (C)	1 kilogram (kg) = 1,000 grams (g)
1 quart = 2 pints (pt)	1 kg = 2.204 lb
1 gallon = 4 quarts (qt)	1 pound (lb) = 16 ounces (oz)
1 gallon = 3.8 liters (L)	
1 cup = 8 fluid ounces (foz)	
1 milliliter = 0.034 fluid ounce (fl oz)	
1 quart = 0.946 liter (L)	

Exercises 31 Convert the measurements.

1) 2.9 m = _____ cm

2) 0.09 liters (L) = _____ mL

3) 17 pounds (lb) = _____ g

4) 273 grams (g) = _____ pounds

5) 368 cm = _____ m

6) 93 ounces (oz) = _____ lb

7) 3.5 feet (ft) = _____ inches (in.)

8) 638 pounds (lb) = _____ kg

9) 1.2 lb = _____ ounces (oz)

10) 0.026 kL = _____ liters (L)

Exercises 32   Convert the measurements.

1) $5\frac{1}{4}$ in. = _____ feet (ft)

2) 573 quarts = _____ pints

3) $2\frac{1}{4}$ liters (L) = _____ mL

4) 4.4 cups = _____ foz

5) 93 pints = _____ cups

6) 182 gallons (gal) = _____ pints

7) 529 mL = _____ liters (L)

8) 51 yards (yd.)= _____ feet (ft)

9) 269 quarts = _____ pints

10) $2\frac{1}{4}$ yards (yd.) = _____ in.

11) 38 yards (yd.) = _____ feet

12) 79 in. = _____ feet (ft)

13) 663 quarts = _____ gallons

14) 0.8 tons (T) = _____ kg

15) $7\frac{3}{5}$ cm = _____ inches

16) 76322 mL = _____ liters (L)

17) $1\frac{1}{2}$ qts = _____ pts

18) 18 yd = _____ ft

Exercises 33 Convert the measurements.

1) 0.52 kg = _____ pounds (lb)

2) 0.73 lb = _____ounces (oz)

3) 623 pints = _____ quarts

4) 0.01 kg = _____ g

5) 827 g = _____ kg

6) $3\frac{1}{4}$ feet (ft) = _____ cm

7) 0.75 feet (ft) = _____ inches (in.)

8) 0.39 yards (yd.) = _____ inches

* Solving Problems

Exercises 34 Solve each problem using the given information.

* Use the following information for Exercises **1-3** below.
   Luke receives an allowance of $5.00 per week from his parents.

1) If his allowance is increased by $0.50, how many weeks will it take for him to amass $31.00?

2) If his allowance is increased by $0.75, how much money will he have at the end of 32 weeks?

3) If his allowance increases by $1.25, how many more weeks will it take before he gets $24.75?

*   Use the following information for Questions **1-2**. A candle is 9 inches long and is completely melted down in 4 hours and 30 minutes.

1. If the candle melts down to $2\frac{1}{4}$ inches, how much time has passed?

    A. 67.5 minutes   **B.** 30 minutes
    C. 1h30 minutes   **D.** 2h

2. If 1 hour 20 minutes passes, how many inches did the candle melt down to?

    A. $2\frac{2}{3}$ **B.** $\frac{2}{3}$

    C. $1\frac{2}{3}$ **D.** $1\frac{1}{4}$

*   Use the following information for Questions **3-4**. The public swimming pools are 50 m long and 25 m wide in the shape of a rectangle and 1.35 m deep. It takes 1 hour 20 minutes to fill an empty pool with water.

3. What is the volume of the swimming pool?

    A. 33.75 m^3 **B.** 67.5 m^3
    C. 1250 m^3 **D.** 1687.5 m^3

4. What is the volume of the water in the pool after it has been filling for 20 minutes?

    A. 421.875 m^3 **B.** 632.8 m^3
    C. 843.75 m^3 **D.** 1265.625 m^3

*   Use the following information for Questions **5-6**. Cowboys Stadium has a retractable dome roof that features an opening that is 125 m long. It takes 12 minutes to open and close the roof.

5. If the dome roof is opened 75 m long, how much longer will it take to open the roof?

    A. 3.6 minutes   **B.** 5.5 minutes
    C. 7.2 minutes   **D.** 14.2 minutes

6. What is the distance of the gap after 5 minutes?

    A. 126.3 m   **B.** 52.6 m
    C. 105.3 m   **D.** 79.2 m

*   Use the following information for Exercises **7-10**. Kevin can type 48 words per minute and takes 28 minutes to type 3 pages.

7. How long will it take to type 360 words?

A. 6 minutes
C. 7 minutes

B. 6.5 minutes
D. 7.5 minutes

8. How long will it take to type one page?

A. ≈9 minutes
C. ≈10 minutes

B. ≈7 minutes
D. ≈12 minutes

9. If he types the same speed for 2 pages, how long will he take?

A. ≈6 minutes
C. ≈8 minutes

B. ≈7 minutes
D. ≈9 minutes

10. How many pages will he type in 2 hours 10 minutes?

A. 7
C. 14

B. 10
D. 21

*   Use the following information for Questions **11-14**. One of the longest bridges in the world is the Manchaca Swamp Bridge, which is 22.81 miles long.

11. If you drive on it at a constant speed of 58 miles per hour, how long will it take to cross the bridge?

A. 11.8
C. 13.8

B. 12
D. 14.8

12. How many more miles do you need to drive after you drove for 15 minutes at 60 mph from the start?

A. 5.8 miles
C. 15 miles

B. 7.81 miles
D. 2.8 miles

13. If you are driving at 50 mph and dove for 14 miles from the start, how many more minutes will it take to cross the bridge?

A. 1.8 minutes
C. 3.6 minutes

B. 2.5 minutes
D. 7.2 minutes

14. What was your speed if you crossed the bridge in 30.4 minutes?

A. 35 mph
C. 45 mph

B. 40 mph
D. 50 mph

* Use the following information for Questions **15-18**. David bought a computer at an electronics store and paid $\frac{3}{4}$ of the price. He paid for the rest by sending \$16.60 every month to the store. The total price of the computer is \$1660.00.

15. How much money did he initially pay for the computer?

    A. \$622.50   **B.** \$822.50
    C. \$1245.00   **D.** \$1445.00

16. How many months will it take until he fully pays for the computer?

    A. 15 months   **B.** 18 months
    C. 20 months   **D.** 25 months

17. How much money does he pay for 7 months?

    A. \$58.10   **B.** \$116.20
    C. \$174.30   **D.** \$232.40

18. If he paid \$265.60, how many months have passed?

    A. 16 months   **B.** 15 months
    C. 14 months   **D.** 13 months

* Use the following information for Questions **19-22**. Jayden is boiling water in a thick glass container for a science experiment. He found that for every 5 minutes, the temperature increases by 10°F. The initial temperature of the water was 75°F.

19. How much time has passed if the temperature of water is 125°F?

    A. 2 minutes   **B.** 3 minutes
    C. 4 minutes   **D.** 5 minutes

20. What is the temperature of water after 38 minutes?

    A. 76°F   **B.** 128°F
    C. 142°F   **D.** 151°F

21. If the water is 105°F, how much time passed?

    A. 5 minutes   **B.** 10 minutes
    C. 15 minutes   **D.** 20 minutes

22. If 1 hour has passed, then what is the temperature of the water?

    A. 65°F   **B.** 155°F
    C. 195°F   **D.** 120°F

CHAPTER 4
Graphing Functions

You will learn how to graph functions and plot coordinates, how to use function tables and find their equations.

1. Understanding Coordinate Grids

a) (x, y) represents the coordinates of a point.
b) The x-axis is the horizontal line and the y-axis is the vertical line in a graph.

4–1. Plot A(–3, 2) on a coordinate grid.

a) First, look at A(–3, 2).

(–3, 2)

y-coordinate
x-coordinate

b) i)   Start at point 0 (origin).
ii) Count 3 units to the left along the x-axis.
iii) Then count 2 units up the y-axis.

y means y-axis.

x means x-axis.

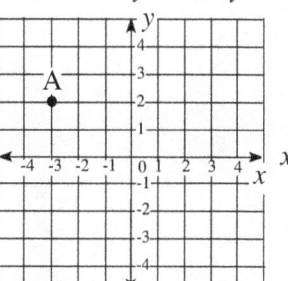

Quick Exercises 1   Plot the given points on a coordinate grid and then connect them.

1) A(–2, –5)

2) B(5, –2)

3) C(2, 6)

4) D(–5, 3)

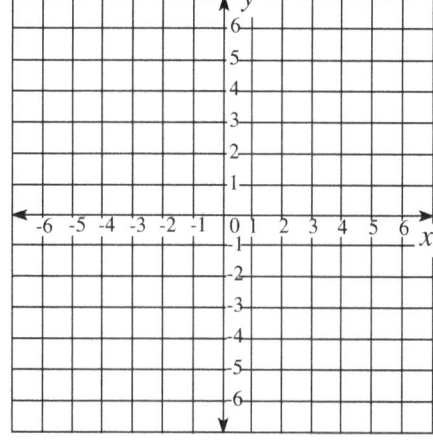

Exercises 1 Plot the given points on a coordinate grid.

1) P(4, 4)

2) Q(2, −3)

3) R(−5, −4)

4) S(−3, 5)

5) T(2, 3)

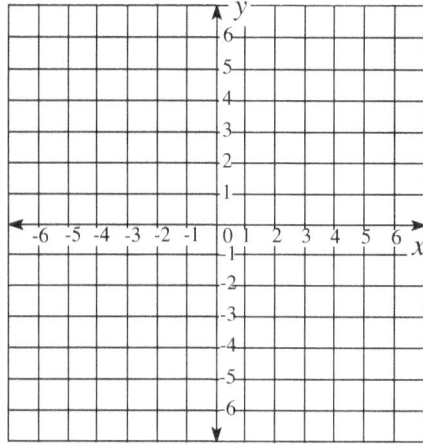

Exercises 2 Find the coordinates of each point.

1) A   2) B

3) C   4) D

5) E   6) F

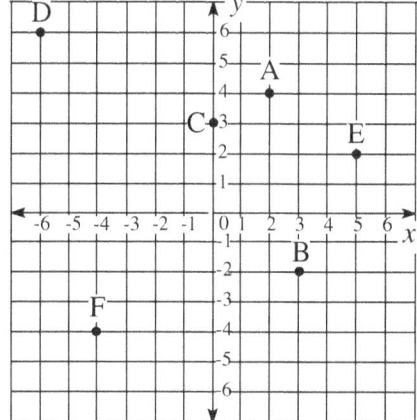

Exercises 3 Use the graph below in order to answer the following questions.

1) Which point is located at (2, 6)?

2) What are the coordinates for C?

3) Which point is at (1, 2)?

4) What are the coordinates for D?

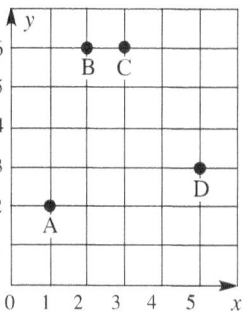

Exercises 4 Use the graph below in order to answer the following questions.

1) Draw A (4, 1), B (6, 3), and C (8, 5). Find the unknown *y*-coordinate of D (10, ?) so that D is on the same line as the other three points.

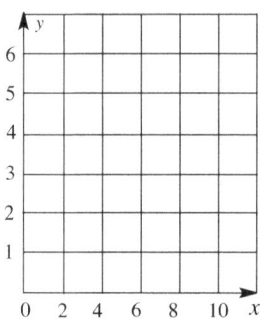

2) Draw A(1, 10) and B(3, 20). Find the unknown *x*-coordinate of C(?, 30) so that C is on the same line as the other two points.

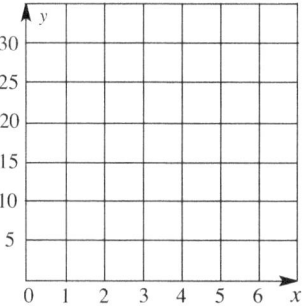

3) Draw A(12, 8) and B(20, 6). Find the unknown coordinates of C(?, ?) and D(?, ?) so that they are on the same line as A and B.

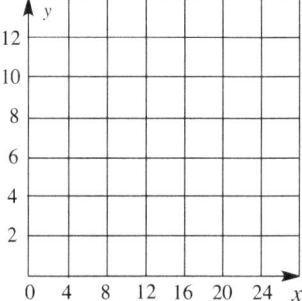

4) Jenny wants to buy donuts. If she spent $3.00 to spend, how many donuts can she buy? Use the table and the graph to determine the total cost of purchasing the donuts.

Number of donuts	0	4	16
Cost	$0	$1.00	$4.00

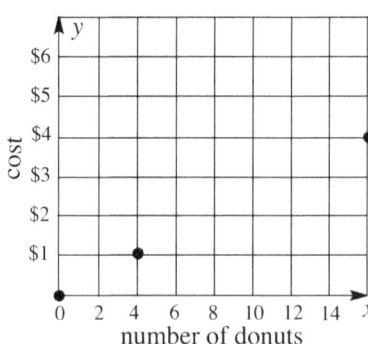

4-2. Find the distance between P and Q on the coordinate grid.

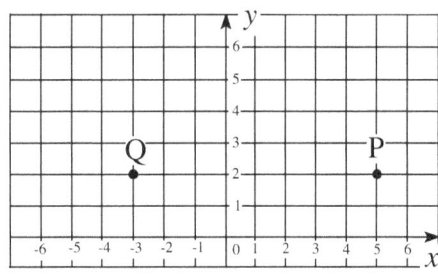

SOLUTION

You can use two ways to find the distance between P and Q on the coordinate grid.
a) First, look at the coordinates of P and Q.

  P(-3, 2) and Q(5, 2)

i) When looking at P(-3, **2**) and Q(5, **2**), you can see that they have the same *y*-coordinates.

ii) Therefore, only be concerned with the *x*-coordinates of P(**-3**, 2) and Q(**5**, 2). Subtract the values of the *x*-coordinates between P and Q or **5** to **-3**, which will be the distance between the two points. Write that $\overline{PQ} = 5 - (-3) = 5 + 3 = 8$. Therefore, the distance between P(-3, 2) and Q(5, 2) is 8 units.

b) Use a graph to count the units between the points.

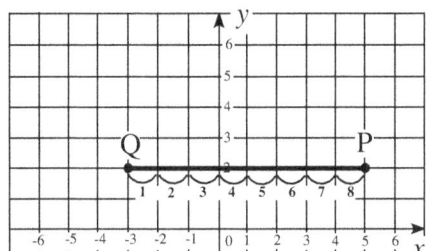

Therefore, the number of units between points P(5, 2) and Q(-3, 2) on the coordinate grid is 8 units.

Quick Exercises 2   Use the graph below.

1) What are the coordinates for A and B?

2) What is the distance between A and B?
Explain your answer.

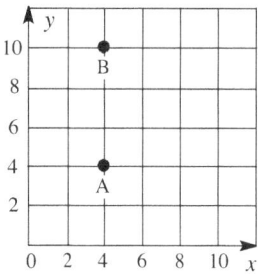

Exercises 5 Use the graph below.

1) What are the coordinates for C and D?

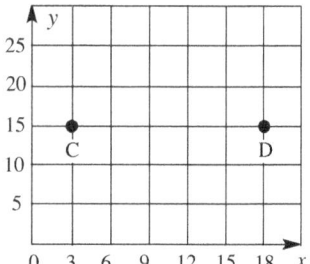

2) What is the distance between C and D? Explain your answer.

3) Name the point that is 8 units away from (6, 2).

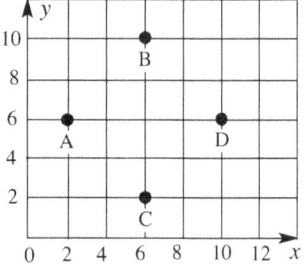

4) Which point can be moved from (6, 10) to the point (10, 6)? Explain your answer.

Exercises 6 Find the distance between the points on each coordinate grid.

1)

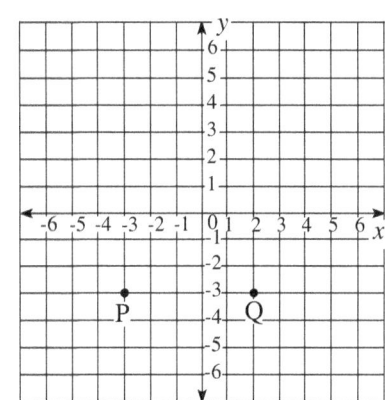

2)

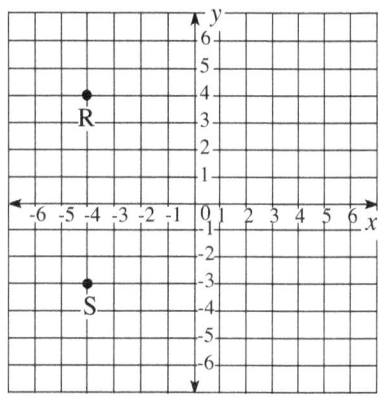

3)

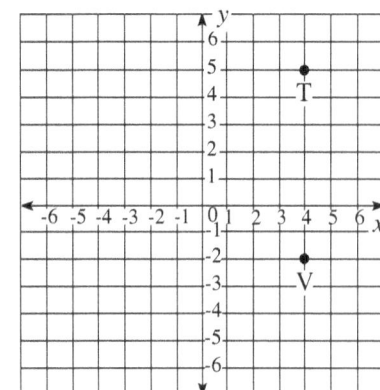

4)

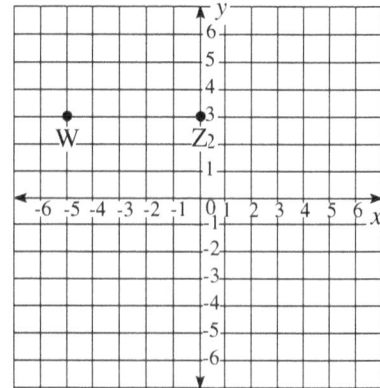

Exercises 7 Graph the coordinates and find the distance between each set of points.

1) R(–3, 3) and E(5, 3)

2) E(5, 3) and C(5, –4)

3) C(5, –4) and T(–3, –4)

4) T(–3, –4) and R(–3, 3)

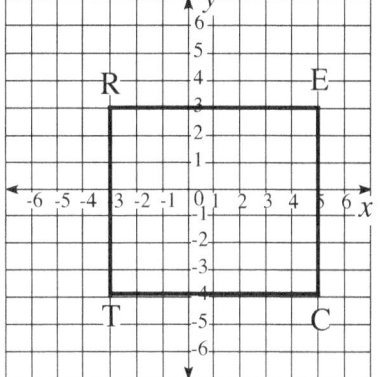

Exercises 8 Graph the coordinates and find the distance between each set of points.

1) A(5, 3) and L(5, –2)

2) L(5, –2) and G(–5, –2)

3) G(–5, –2) and B(–5, 3)

4) B(–5, 3) and A(5, 3)

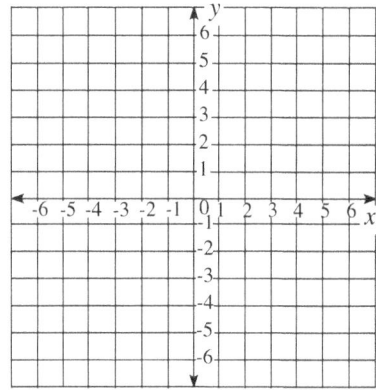

Exercises 9 Use the graph to complete the function table.

	A	B	C	D	E
x	–3				
y			–1		

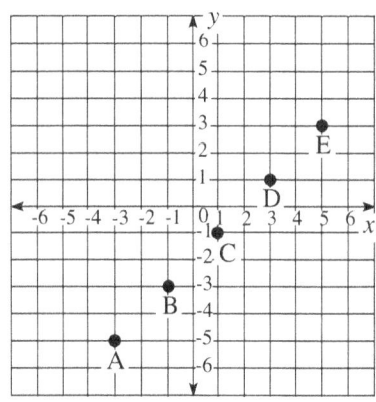

* For Questions **1-4**, use the graph.

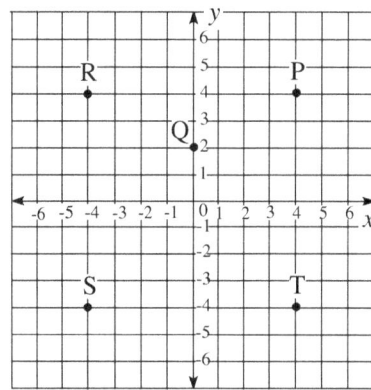

1. Which of the following are the coordinates for P?

    A. (4, 4)   **B.** (4, −4)
    C. (−4, −4)   **D.** (−4, 4)

2. Which of the following points is located at (−4, 4)?

    A. P   **B.** R
    C. S   **D.** T

3. Which of the following coordinates is located at T?

    A. (4, 4)   **B.** (4, −4)
    C. (−4, −4)   **D.** (−4, 4)

4. Which of the following points is located at (−4, −4)?

    A. Q   **B.** R
    C. S   **D.** T

* For Questions **5-6**, use the graph to find each location.

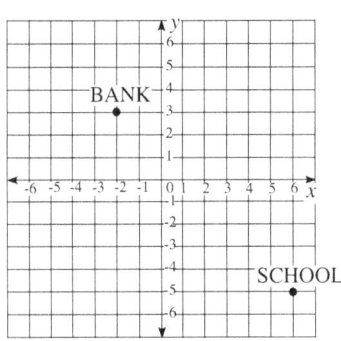

5. Which of the following are the coordinates for the bank?

       A. $(-5, 6)$ **B.** $(6, -5)$

       C. $(-2, 3)$ **D**. $(3, -2)$

6. Which of the following are the coordinates for the school?

       A. $(6, -5)$ **B.** $(-5, 6)$

       C. $(5, 6)$ **D**. $(6, 5)$

*   For Questions **7-10**, use the graph to find each location.

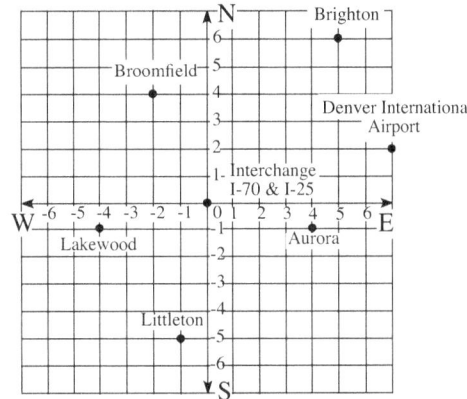

7. What are the coordinates of the Denver National Airport?

       A. $(7, 2)$ **B.** $(2, 7)$

       C. $(-1, 5)$ **D**. $(5, -1)$

8. Which of the following places is located at $(-1, -5)$?

       A. Littleton   **B.** Broomfield

       C. Aurora   **D**. Lakewood

9. What are the coordinates of Broomfield?

    A. (−1, 4)   **B.** (4, −2)

    C. (−2, 4)   **D.** (2, 4)

10. Nick's family is touring the city of Denver. If they are driving their car from the Denver International Airport and head 7 units west and 2 units south, what place are they heading towards?

    A. Brighton   **B.** Broomfield

    C. Aurora   **D.** Littleton

* For Questions **11-14**, use the graph.

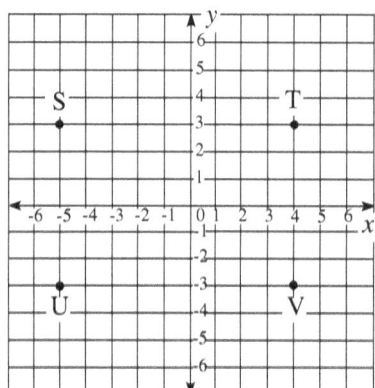

11. Which of the following are the coordinates for S?

    A. (−5, 3)   **B.** (−5, −3)

    C. (4, 3)   **D.** (3, 4)

12. Which of the following points is located at (4, −3)?

    A. S   **B.** T

    C. U   **D.** V

13. What is the distance between S and T?

    A. 4 units   **B.** 7 units

    C. 9 units   **D.** 10 units

14. What is the distance between S and U?

    A. 4 units   **B.** 6 units

    C. 9 units   **D.** 10 units

*   For Questions **15-20**, use the graph.

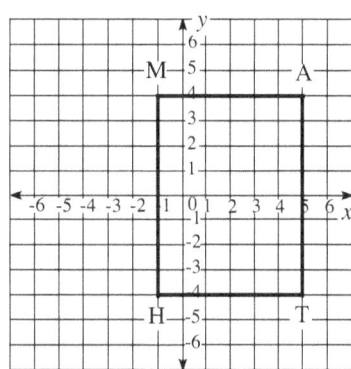

15. Which of the following coordinates is located at A?

     A. (5, –4)   **B.** (–4, 5)

     C. (5, 4)   **D**. (4, 5)

16. Which of the following points is located at (–1, –4)?

     A. M   **B.** A

     C. T   **D**. H

17. What is the distance of $\overline{MA}$?

     A. 1 unit   **B.** 4 units

     C. 6 units   **D**. 10 units

18. What is the distance of $\overline{AT}$?

     A. 1 unit   **B.** 4 units

     C. 6 units   **D**. 8 units

19. Find the distance between (–1, –4) and (5, –4).

     A. 1 unit   **B.** 4 units

     C. 6 units   **D**. 10 units

20. Given the two points located at (5, 4) and (5, –4), what is the name of the segment they form?

     A. $\overline{MA}$ **B.** $\overline{MH}$

     C. $\overline{AT}$ **D**. $\overline{HT}$

2. Function Table

4–3. What is a function table?

> Function table: A function table can be organized into the table to write the value of an output for each input.

4–3. What is the unknown value on the function table below?

x	−1	0	1	2
y	−5	−1	?	7

SOLUTION

Find the intervals between the numbers shown in a function table.

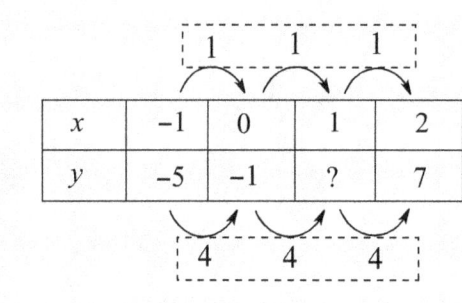

The interval between the x-values is 1, which means that each value of x increases by 1.
The interval between the y-values is 4, which means that each value of y increases by 4.

So the next value of y is 3.
$$4 + (-1) = 3$$

Quick Exercises 3   Find the unknown value of each function table.

1)

x	4	5	7	9
y	11	?	20	26

2)

x	−3	0	3	6
y	−8	?	−2	1

3)

x	−1	2	5	8
y	?	5	11	17

4)

x	−3	0	?	9
y	6	0	−6	?

4-4. Use the given equation to make a function table.

$$y = 2 \times x$$

SOLUTION

Find the value for y.
$y = 2 \times x$ or $y = 2x$ (The sign ($\times$) can be omitted between the number and the variable).

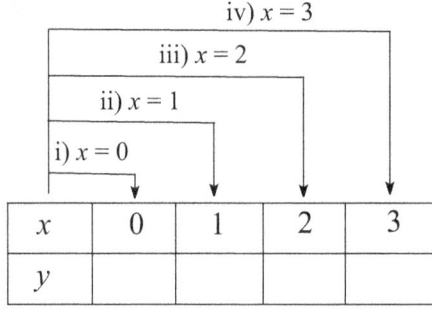

i) If $x = 0$, then $y = 0$.

$y = 2 \times x$	Original equation
$= 2 \times 0$	Substitute 0 for x.
$= 0$	

ii) If $x = 1$, then $y = 2$.

$y = 2 \times x$	Original equation
$= 2 \times 1$	Substitute 1 for x.
$= 2$	

iii) If $x = 2$, then $y = 4$.

$y = 2 \times x$	Original equation
$= 2 \times 2$	Substitute 2 for x.
$= 4$	

iv) If $x = 3$, then $y = 6$.

$y = 2 \times x$	Original equation
$= 2 \times 3$	Substitute 3 for x.
$= 6$	

$y = 2 \times x$

x	0	1	2	3
y	0	2	4	6

Quick Exercises 4   Complete the function table for each equation.

1) $y = 2x$

x	-4	-1	2	5
y				

2) $y = 2x$

x				
y	-4	0	4	6

3) $y = -x$

x	-2	-1	0	1
y				

4) $y = -x$

x				
y	5	1	-3	-11

4–5. Use the given equation to make a function table.

$$y = 3x + 1$$

SOLUTION

Find the value for y.
$$y = 3x + 1$$

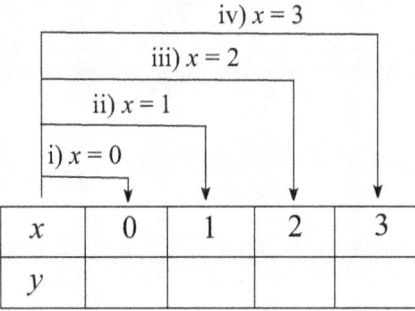

i)   If $x = 0$, then $y = 1$.
$$y = 3x + 1 \qquad \text{Original equation}$$
$$= 3(0) + 1 \qquad \text{Substitute 0 for } x.$$
$$= 1$$

ii) If $x = 1$, then $y = 4$.
$$y = 3x + 1 \qquad \text{Original equation}$$
$$= 3(1) + 1 \qquad \text{Substitute 1 for } x.$$
$$= 4$$

iii) If $x = 2$, then $y = 7$.
$$y = 3x + 1 \qquad \text{Original equation}$$
$$= 3(2) + 1 \qquad \text{Substitute 2 for } x.$$
$$= 7$$

iv) If $x = 3$, then $y = 10$.
$$y = 3x + 1 \qquad \text{Original equation}$$
$$= 3(3) + 1 \qquad \text{Substitute 3 for } x.$$
$$= 10$$

$\Downarrow$ $y = 3x + 1$

x	0	1	2	3
y	1	4	7	10

Quick Exercises 5   Complete the function table for each equation.

1) $y = 3x + 1$

x	−4	−1	2	5
y				

2) $y = 3x + 1$

x				
y	−5	1	7	16

3) $y = -x + 1$

x	−4	−1	2	5
y				

4) $y = -2x - 1$

x				
y	−5	1	7	16

4-6. Use the given equation to make a function table.

$$y = x \div 2$$

SOLUTION

a) It is given that x divided by 2 equals y. You can rewrite the equation so that y multiplied by 2 equals x.

$$y = x \div 2 \text{ or } y = \frac{x}{2} \qquad \text{Original Equation}$$

$$2 \times y = \frac{x}{2} \times 2 \qquad \text{Multiply each side by 2.}$$

$$2 \times y = x \qquad \text{Simplify.}$$

So, you can use this equation to find the function table or you can use another way as shown in the following.

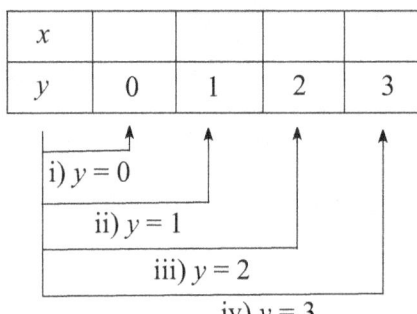

x				
y	0	1	2	3

i) $y = 0$

ii) $y = 1$

iii) $y = 2$

iv) $y = 3$

$y = x \div 2$ or
$x = 2y$

x	0	2	4	6
y	0	1	2	3

i) If $y = 0$, then $x = 0$.
$\quad x = 2y \qquad$ Original equation
$\quad\quad = 2(0) \qquad$ Substitute 0 for y.
$\quad\quad = 0$

ii) If $y = 1$, then $x = 2$.
$\quad x = 2y \qquad$ Original equation
$\quad\quad = 2(1) \qquad$ Substitute 1 for y.
$\quad\quad = 2$

iii) If $y = 2$, then $x = 4$.
$\quad x = 2y \qquad$ Original equation
$\quad\quad = 2(2) \qquad$ Substitute 2 for y.
$\quad\quad = 4$

iv) If $y = 3$, then $x = 6$.
$\quad x = 2y \qquad$ Original equation
$\quad\quad = 2(3) \qquad$ Substitute 3 for y.
$\quad\quad = 6$

b) Check the solution.
   Once you make the table for the equation $x = 2y$, check your solution by substituting $x = 0, 2, 4,$ and 6 in the original equation of $y = x \div 2$.

Quick Exercises 6   Complete the function table for each equation.

1) $y = \dfrac{x}{2}$

x				
y	−1	0	1	2

2) $y = \dfrac{1}{3}x + 1$

x				
y	0	1	2	3

Exercises 10 Find the unknown value in each function table.

1)

x	0	1	2	3
y	?	2	4	6

2)

x	−3	?	3	6
y	1	2	3	4

3)

x	−2	−1	0	1
y	?	−4	1	6

4)

x	−2	−1	0	6
y	−7	−4	−1	?

5)

x	−3	0	3	6
y	3	?	13	18

6)

x	−1	0	1	?
y	2	4	6	12

Exercises 11 Use the given equation to make a function table.

1) $y = 4 \times x$

x	0	1	2	3
y				

2) $y = 3 \times x$

x	−1	0	1	2
y				

3) $y = 2 \times x$

x	−1	0	1	2
y				

4) $y = x$

x	−3	0	3	6
y				

5) $y = 5 \times x$

x	0	1	2	3
y				

6) $y = 5x$

x				
y	−10	−5	0	5

7) $y = -3x$

x	−3	0	3	6
y				

8) $y = -4x$

x				
y	0	8	16	24

Exercises 12 Use the given equations to make a function table.

1) $y = 3x - 2$

x	0	1	2	3
y				

2) $y = 2x + 1$

x	−1	0	1	2
y				

3) $y = 2x - 1$

x	−1	0	1	2
y				

4) $y = 3x - 1$

x	−3	0	3	6
y				

Exercises 13 Use the given equations to make a function table.

1) $y = x - 1$

x	0	1	2	3
y				

2) $y = -x + 1$

x	0	1	2	3
y				

3) $y = -4x + 1$

x	0	1	2	3
y				

4) $y = 4x - 2$

x				
y	−6	2	10	18

5) $y = -3x + 1$

x	−3	0	3	6
y				

6) $y = 3x - 2$

x				
y	−2	1	4	7

Exercises 14 Use each given equation to make a function table.

1) $y = \dfrac{1}{2}x$

x				
y	−3	−1	1	9

2) $y = \dfrac{1}{3}x$

x				
y	−2	−1	0	5

3) $y = -\dfrac{1}{3}x$

x				
y	−2	−1	0	5

4) $y = \dfrac{1}{2}x - 1$

x				
y	−3	−1	1	9

Exercises 15 Use each given equation to make a function table.

1) $y = -\dfrac{1}{2}x + 1$

x				
y	−2	−1	0	5

2) $2(y + 1) = -x$

x				
y	−3	−1	1	9

3) $y = -\dfrac{1}{3}x - 1$

x				
y	−2	−1	0	5

4) $y = \dfrac{1}{3}x + 1$

x				
y	−2	−1	0	5

5) $y = \dfrac{1}{3}x + 2$

x				
y	−2	−1	0	5

6) $y = -\dfrac{1}{3}x + 2$

x				
y	−2	−1	0	5

1. What is the unknown value on the function table below?

x	3	4	5	6
y	10		16	19

A. 11   B. 12
C. 13   D. 14

2. What is the unknown value on the function table below?

x	7	9	11	12
y	20		32	35

A. 22   B. 25
C. 26   D. 29

3. What is the unknown value on the function table below?

x	0	1	2	3
y	5	−1	?	−13

A. −5   B. −6
C. −7   D. −8

4. What is the unknown value on the function table below?

x	−3	0	3	9
y	10	1	−8	

A. −16   B. −17
C. −25   D. −26

5. What is the unknown value on the function table below?

x	−2	−1	0	1
y	−5	−2		4

A. −1   **B.** 0
C. 1   **D.** 2

6. Use the function table below. If the value of x is 15, what is the value of y?

x	0	3	4
y	4	10	12

A. 24   **B.** 26
C. 30   **D.** 34

7. Use the function table below. If the value of x is 2, what is the value of y?

x	5	7	9
y	6	10	14

A. 0   **B.** 1
C. 2   **D.** 4

8. What is the unknown value of the function table below?

x	−3		3	12
y	−1	0	1	4

A. −2   **B.** −1
C. 0   **D.** 1

9. What is the unknown value of the function table below?

x	0	3	6	12
y	−1	0	1	

A. 2   **B.** 3
C. 4   **D.** 5

10. Use the function table below. Given the value of y is 6, what is the value of x?

x	2	4	6
y	2	3	4

A. 7

C. 9

B. 8

D. 10

11. Use the function table below. If the value of y is 0, what is the value of x?

x	12	18	24
y	5	7	9

A. 0

C. 2

B. 1

D. −3

12. If the value of x is 99, what is the value of y using the equation below?

$$y = \frac{1}{3}x - 1$$

A. 23

C. 32

B. 35

D. 19

13. Which of the following function tables represents the equation $y = x \div 4$?

A.

x	−4	0	4
y	−4	0	4

B.

x	8	12	16
y	2	3	4

C.

x	−4	0	4
y	−1	1	2

D.

x	−4	4	8
y	−1	0	2

14. Which of the following function tables represents the equation $y = 2x - 1$?

A.

x	0	1	2	3
y	−1	0	4	6

B.

x	0	1	2	3
y	0	1	3	5

C.

x	0	1	2	3
y	−1	1	3	5

D.

x	−1	1	3	5
y	0	1	2	3

15. Which of the following function tables represents the equation $y = -4x$?

A.

x	-3	-2	-1	0
y	-12	-8	-4	0

B.

x	-1	0	1	2
y	4	0	-4	-8

C.

x	1	3	5	7
y	-4	-1	1	3

D.

x	-2	-1	0	1
y	-8	-4	0	4

16. Which of the following function tables represents the equation $y = x + 3$?

A.

x	-1	1	3
y	0	2	5

B.

x	-1	0	1
y	2	3	5

C.

x	2	3	4
y	5	6	7

D.

x	-3	0	3
y	-3	0	3

17. Which of the following function tables represents the equation $y = -x - 3$?

A.

x	2	4	7
y	-4	-5	-9

B.

x	-2	0	2
y	-1	-3	-5

C.

x	-1	1	2
y	-4	-3	-5

D.

x	-2	-1	1
y	-1	-4	-3

3. Writing Linear Equations

4–7. Find the equation of each function table.

x	0	1	2	3
y	0	5	10	15

SOLUTION

a) First, look at $x = 0$.
    When $x = 0$, then $y = 0$. ⟵ It is the same as $y = \square x$.
So, you can write the equation as $y = \square x + 0$ because if $x = 0$, then $y = 0$.

b) Next, look for $\square$.
    Check the relationship between the values of x and y.

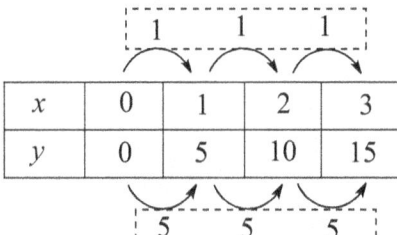

The values of x constantly increase by 1.
The values of y constantly increase by 5.

This means when the values of x increase by **1**, then the values of y increase by **5**. So, the equation of this table is **1**$y =$ **5**x or $y = 5x$.

Quick Exercises 7　　Find the equation of each function table.

1)

x	−2	−1	0	1
y	−4	−2	0	2

2)

x	−4	−2	0	2
y	−12	−6	0	6

3)

x	−5	−3	−1	1
y	5	3	1	−1

4–8. Find the equation of each function table.

x	0	1	2	3
y	1	6	11	16

SOLUTION

a) First, look at x = 0.
   When x = 0, then y = 1.

x	**0**	1	2	3
y	**1**	6	11	16

So, you can write the equation as $y = \square x + 1$ because when x = 0, then y = 1.

b) Next, check the relationship between the values of x and y.

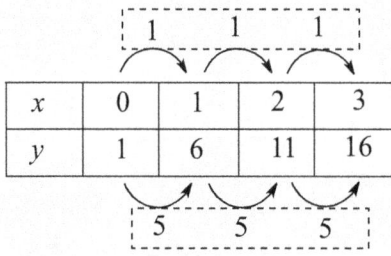

This means when the values of x increase by 1, then the values of y increase by **5**. So, the equation of the information on this table is $y = \mathbf{5}x + 1$.

Quick Exercises 8   Find the equation of each function table.

1)

x	−4	−2	0	2
y	−10	−4	2	8

2)

x	−2	−1	0	1
y	−3	−1	1	3

3)

x	−5	−3	−1	1
y	7	5	3	1

4-9. Find the equation of each function table.

x	-2	-1	0	1
y	-7	-4	-1	2

SOLUTION

a) First, look at $x = 0$.
    When $x = 0$, then $y = -1$.
    So, you can write the equation that is $y = ? \, x - 1$.

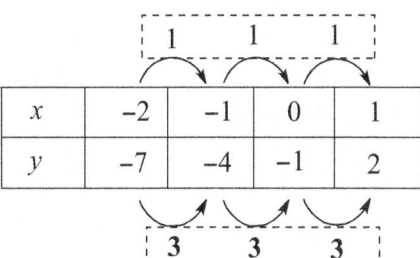

b) Next, find the intervals between the values.
    Check the relationship between the values of x and y.

The values of x constantly increase by 1.
The values of y constantly increase by 3.
Now, you can find the equation, which is $y = 3x - 1$.

Quick Exercises 9   Find the equation of each function table.

1)

x	-4	-2	0	2
y	-14	-8	-2	4

2)

x	-4	-2	0	2
y	-9	-5	-1	3

3)

x	-5	-3	-1	1
y	9	5	1	-3

4–10. Find the equation of the function table.

x	–4	–2	0	2
y	–2	–1	0	1

SOLUTION

a) First, look at $x = 0$.
   When $x = 0$, then $y = 0$.
   So, you can write the equation that is
   $y = ?\,x + \mathbf{0}$ or $y = ?\,x$.

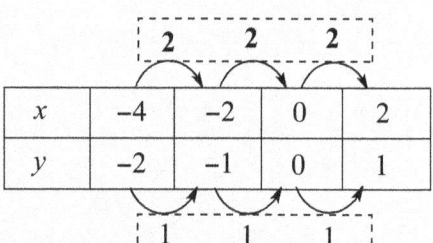

$$y = ?\,x + \mathbf{0}$$

b) Next, find the intervals between the
   values.

   Check the relationship between the
   values of x and y.

The values of x constantly increase by **2**.
The values of y constantly increase by **1**.
Now, you can find the equation, which is
$2y = x$ or $y = \dfrac{1}{2}x$.

Quick Exercises 10   Find the equation of each function table.

1)

x	–6	–3	0	3
y	–2	–1	0	1

2)

x	–9	0	9	18
y	–3	0	3	6

3)

x	–18	–10	–2	6
y	9	5	1	–3

4–11. Find the equation of each function table.

x	–6	–4	–2	0
y	–2	–1	0	1

SOLUTION

a) First, look at $x = 0$.
   When $x = 0$, then $y = 1$.
   So, you can write the equation that is
   $y = ? \, x + 1$.

   $y = ?x + 1$ or $y - 1 = ?x$

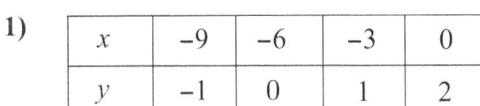

$y = ? \, x + 1$

b) Next, find the intervals between the values.
   Check the relationship between the values of x and y.

The values of x constantly increase by 2.
The values of y constantly increase by 1.
Now, you can find the equation, which is

$2(y - 1) = x$ or $y = \dfrac{1}{2}x + 1$.

Quick Exercises 11   Find the equation of each function table.

1)

x	–9	–6	–3	0
y	–1	0	1	2

2)

x	–6	0	3	6
y	–3	–1	0	1

3)

x	–17	–9	–1	7
y	9	5	1	–3

Exercises 16 Find the equation of each function table.

1)

x	−2	−1	0	1
y	−10	−5	0	5

2)

x	−8	−4	0	4
y	16	8	0	−8

3)

x	−2	0	2	4
y	−12	0	12	24

4)

x	−6	0	6	12
y	−12	0	12	24

Exercises 17 Find the equation of each function table.

1)

x	−4	0	4	8
y	−10	2	14	42

2)

x	−1	0	1	2
y	3	4	5	6

3)

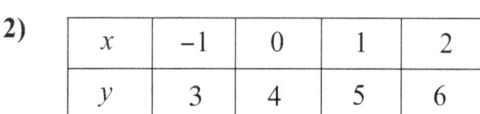

x	2	4	6	8
y	3	7	11	15

4)

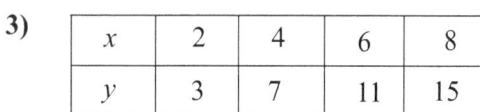

x	−6	−3	0	3
y	4	1	−2	−5

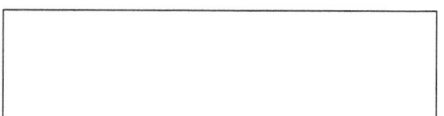

Exercises 18 Find the equation of each function table.

1)

x	−3	−6	−9	−12
y	1	2	3	4

2)

x	−4	0	4	8
y	−1	0	1	2

3)

x	−12	−8	−4	0
y	−6	−4	−2	0

4)

x	6	0	−6	−12
y	−3	0	3	6

Exercises 19 Find the equation of each function table.

1)

x	6	3	0	−3
y	0	1	2	3

2)

x	−12	−8	−4	0
y	−2	−1	0	1

3)

x	0	4	8	12
y	−1	1	3	5

4)

x	2	0	−2	−4
y	−3	−2	−1	0

1. What is the value of (Δ) given the information from the table below?

x	0	1	2	3
y	0	2	4	6

$$y = (\Delta)x$$

A. 1   B. 2
C. 3   D. 4

2. What is the value of (Δ) given the information from the table below?

x	−1	0	1	2
y	−3	0	3	6

$$y = (\Delta)x$$

A. 1   B. 2
C. 3   D. 4

3. What is the value of (Δ) given the information from the table below?

x	0	1	2	3
y	0	1	2	3

$$y = (\Delta)x$$

A. 1   B. 2
C. 3   D. 4

4. What is the value of (Δ) given the information from the table below?

x	0	2	4	6
y	0	6	12	18

$$y = (\Delta)x$$

A. 3   B. 4
C. 5   D. 6

5. What is the value of (Δ) given the information from the table below?

x	2	4	6	8
y	14	28	42	56

$$y = (Δ)x$$

A. 4   **B.** 5
C. 6   **D.** 7

6. Which equation represents the function table below?

x	0	1	2	3
y	2	7	12	17

A. $y = 5 \times x + 2$ **B.** $y = 5x - 1$
C. $y = 5x + 4$ **D.** $y = 5x$

7. Which equation represents the function table below?

x	−1	1	3	5
y	−2	2	6	10

A. $y = 2 \times x + 1$ **B.** $y = x - 1$
C. $y = x + 4$ **D.** $y = 2x$

8. Which equation represents the function table below?

x	−16	−12	−8	−4
y	−1	0	1	2

A. $y = 16x$ **B.** $y = 4x - 12$
C. $x = 4y - 12$ **D.** $x = 3 \times y - 12$

9. Which equation represents the function table below?

x	8	10	12
y	5	6	7

A. $2(y - 1) = x$ **B.** $y = x \div 2 - 1$
C. $2(y + 1) = x$ **D.** $y = 2 \times x - 2$

10. Which equation represents the function table below?

x	0	1	2	3
y	0	4	8	12

A. $y = 3x + 1$ **B.** $y = 5x - 1$
C. $y = x + 3$ **D.** $y = 4x$

11. Which equation represents the function table below?

x	0	2	4	6
y	7	9	11	13

A. $y = x + 1$ **B.** $y = x + 7$
C. $y = 6x + 1$ **D.** $y = 7x + 1$

12. Which equation represents the function table below?

x	0	1	2	3
y	−2	−1	0	1

A. $y = 2x - 2$ **B.** $y = -1x - 1$
C. $y = x - 2$ **D.** $y = 3x - 2$

13. Which equation represents the function table below?

x	0	1	2	3
y	1	7	13	19

A. $y = x + 1$ **B.** $y = x + 7$
C. $y = 6x + 1$ **D.** $y = 7x + 1$

14. Which equation represents the function table below?

x	−1	1	3
y	−2	4	10

A. $y = 2x + 1$ **B.** $y = x - 1$
C. $y = x + 4$ **D.** $y = 3x + 1$

15. Which equation represents the function table below?

x	−1	1	3
y	−5	−3	−1

A. $y = -2 \times x - 4$　　　　　　**B.** $y = x - 4$
C. $y = 2x - 4$　　　　　　　　　**D.** $y = 2 \times x - 3$

16. Which equation represents the function table below?

x	−1	2	3	5
y	−1	8	11	17

A. $y = 2x + 1$　　　　　　　　**B.** $y = 2x - 2$
C. $y = 3x + 2$　　　　　　　　**D.** $y = 4 \times x$

17. Which equation represents the function table below?

x	−1	0	1	2
y	−2	−1	0	1

A. $y = x + 1$　　　　　　　　**B.** $y = x - 1$
C. $y = 2x + 3$　　　　　　　　**D.** $y = 2 \times x + 4$

18. Which of the following function tables represent the equation $y = x + 3$?

A.

x	−3	0	3
y	−3	0	3

B.

x	2	3	4
y	6	7	8

C.

x	−1	0	1
y	2	3	4

D.

x	−1	1	3
y	0	2	5

19. Which equation represents the function table below?

x	−2	−1	0	1
y	−9	−5	−1	3

A. $y = 2x + 1$ **B.** $y = 4x - 2$
C. $y = 3x + 2$ **D.** $y = 4 \times x$

20. Which equation represents the function table below?

x	4	5	7	9
y	11	14	20	26

A. $y = 3 \times x + 1$ **B.** $y = 3x - 1$
C. $y = 2x + 3$ **D.** $y = 2 \times x + 4$

21. Which of the following function tables represent the equation $y = 4x + 1$?

A.

x	2	4	7
y	8	16	14

B.

x	−2	0	2
y	−8	1	8

C.

x	−1	1	2
y	−3	5	9

D.

x	−2	−1	1
y	−9	−3	5

22. Which equation represents the function table below?

x	10	5	0	−5
y	−2	−1	0	1

A. $x = -\dfrac{1}{5}y$ **B.** $x = \dfrac{1}{5}y$
C. $y = \dfrac{1}{5}x$ **D.** $y = -\dfrac{1}{5}x$

23. Which equation represents the function table below?

x	8	4	0	−4
y	−3	−1	1	3

A. $y = \dfrac{1}{2}x - 1$ **B.** $y = \dfrac{1}{2}x + 1$
C. $y = -\dfrac{1}{2}x - 1$ **D.** $y = -\dfrac{1}{2}x + 1$

24. Which equation represents the function table below?

x	-12	-6	0	6
y	-6	-3	0	3

A. $x = -\dfrac{1}{2}y$

B. $x = \dfrac{1}{2}y$

C. $y = \dfrac{1}{2}x$

D. $y = -\dfrac{1}{2}x$

25. Which equation represents the function table below?

x	10	4	2	0
y	-6	-3	-2	-1

A. $y = \dfrac{1}{2}x - 1$

B. $y = \dfrac{1}{2}x + 1$

C. $y = -\dfrac{1}{2}x - 1$

D. $y = -\dfrac{1}{2}x + 1$

26. Which equation represents the function table below?

x	-6	0	9	18
y	-3	-1	2	5

A. $3(y + 1) = x$

B. $3(y - 1) = x$

C. $y = \dfrac{x}{3} + 1$

D. $y = -\dfrac{1}{3}x - 1$

27. Which equation represents the function table below?

x	-8	-4	0	4
y	-25	-13	-1	11

A. $y = 3 \times x + 1$

B. $y = 3x - 2$

C. $y = 2x + 3$

D. $y = 2 \times x + 4$

CHAPTER 5
Graphing Linear Equations

You will learn how to graph linear equations and how to find derive linear equations from function tables, and also deal with parallel and perpendicular lines.

1. Graphing Relations and Function Table

5–1. Use the table to draw the graph and explain the table.

x	−4	−2	0	2
y	−4	−2	0	2

SOLUTION

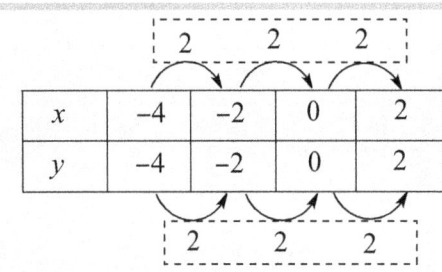

When $x = 0$, $y = 0$.
The values of x constantly increase by **2**.
The values of y constantly increase by **2**.

So the equation in the function table shows $2y = 2x$ or $y = x$.
The values of x and y are equal to each other.

a) The table gives the coordinates of x and y for $y = x$.

b) Plot the coordinates on the grid.

c) The coordinates given determines the relationship between x and y. Plot the points on the graph.

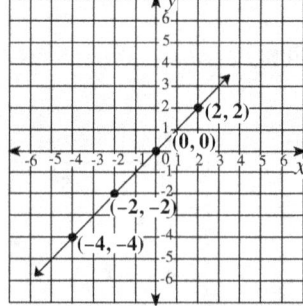

<u>Quick Exercises</u> 1   Use the table to draw the graph and explain the table.

x	-3	-1	1	3
y	9	3	-3	-9

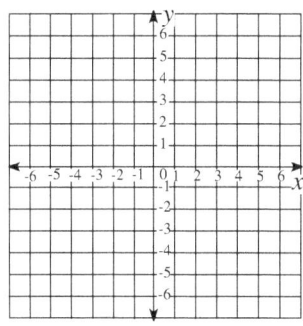

5-2. Use the table to draw a graph and explain the table.

x	-4	-2	0	2
y	-3	-1	1	3

SOLUTION

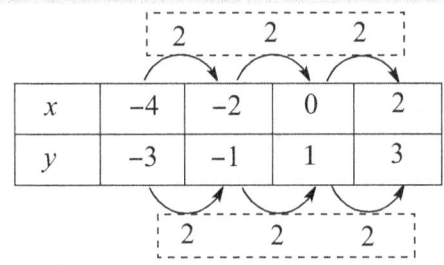

When $x = 0$, $y = 1$.
The values of x constantly increase by **2**.
The values of y constantly increase by **2**.

So you can find the equation, which is **2**(y − 1) = **2**x or $y = x + 1$.
The values of x and y increases by the same amount but the values of y are higher by 1.

a) The table gives the coordinates of x and y for the equation $y = x + 1$.

b) Plot the coordinates on the grid.

c) The coordinates given determines the relationship between x and y. Plot the points on the graph.

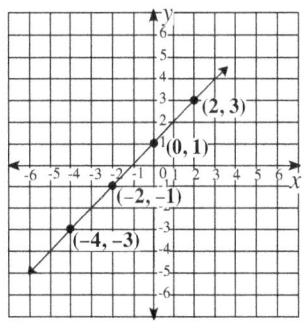

Quick Exercises 2   Use the function table to find the equation, then graph it.

x	−4	−2	0	2
y	−2	0	2	4

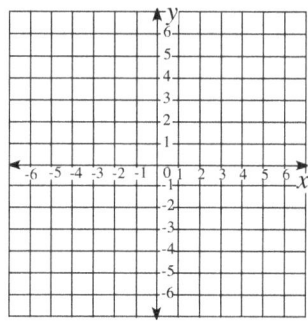

5–3. Use the graph to make a function table and equation. Explain the table and graph.

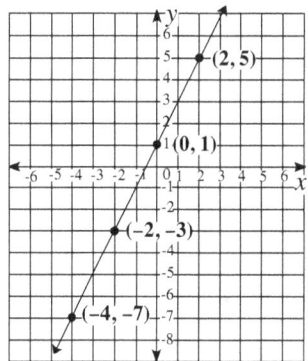

SOLUTION

	2		2		2	
x	−4	−2	0	2
y	−7	−3	1	5
	4		4		4	

The values of x and y increase by 2 and 4 respectively. So the equation of the line is $2(y - 1) = 4x$ or $y = 2x + 1$.
The values of y shift up by 1 from the origin $(0, 0)$.

Quick Exercises 3   Use the graph to make a function table and equation. Explain the table and graph.

1)
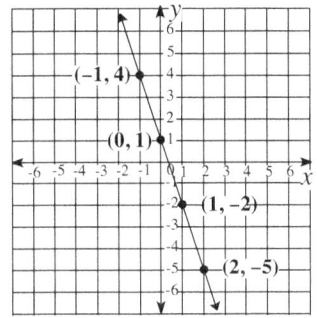

x				
y				

2)
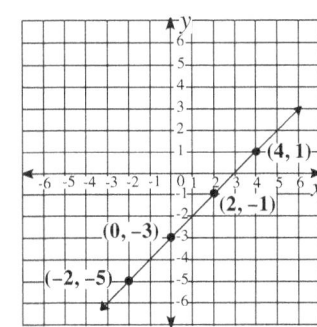

x				
y				

Exercises 1   Use the tables to draw a graph. Explain the table and graph.

1)

x	5	6	7	8
y	15	18	21	24

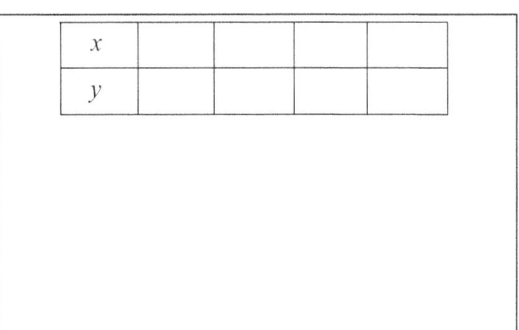

2)

x	−4	−2	0	2
y	−9	−5	−1	3

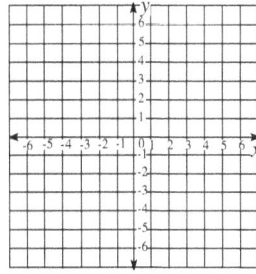

Exercises 2 Use the function table to draw a graph. Explain the table and graph.

1)

x	−3	0	3	6
y	−1	0	1	2

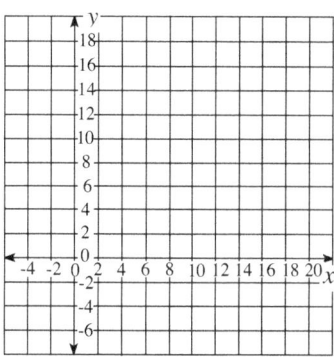

2)

x	4	8	12	16
y	2	4	6	8

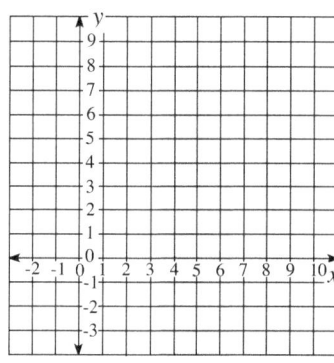

3)

x	2	3	4	5
y	5	7	9	11

4)

x	−2	0	2	5
y	−5	−1	3	9

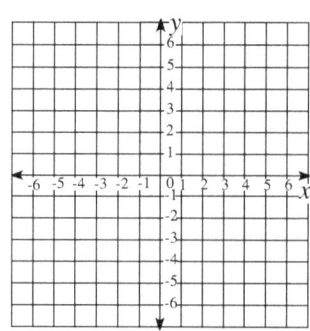

Exercises 3 Use the function table to draw a graph. Explain the table and graph.

1)

x	−4	−2	0	2
y	7	3	−1	−5

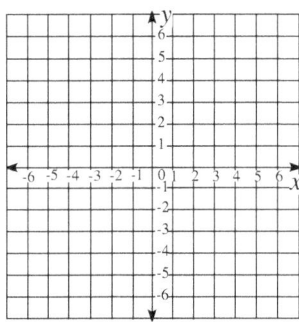

2)

x	−3	0	3	6
y	−8	1	10	19

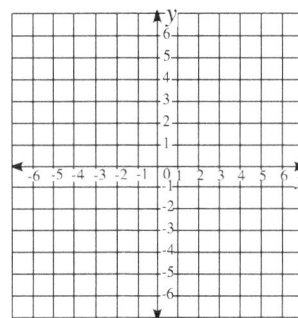

3)

x	−4	0	4	8
y	−3	−1	1	3

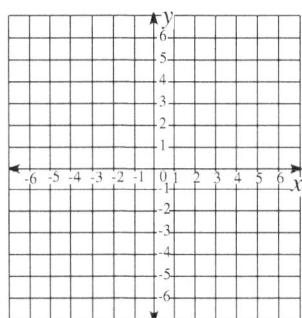

4)

x	18	12	6	0
y	−4	−2	0	2

Exercises 4 Use the tables to draw a graph. Explain the table and graph.

1)

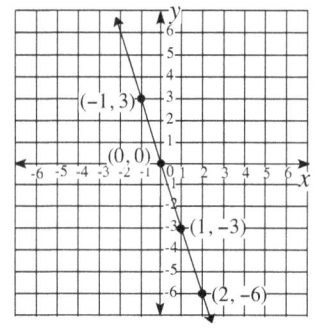

x				
y				

2)

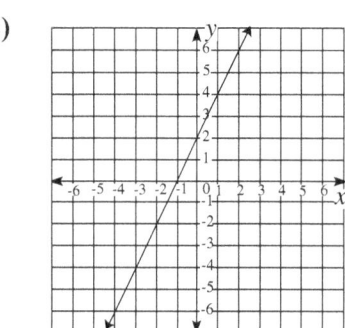

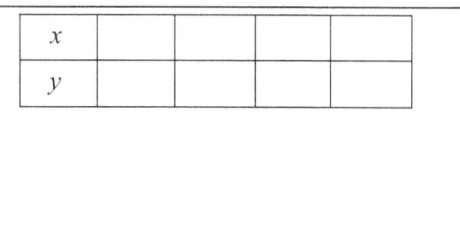

3)

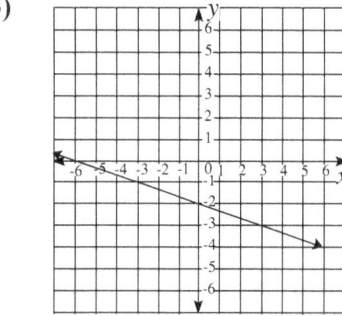

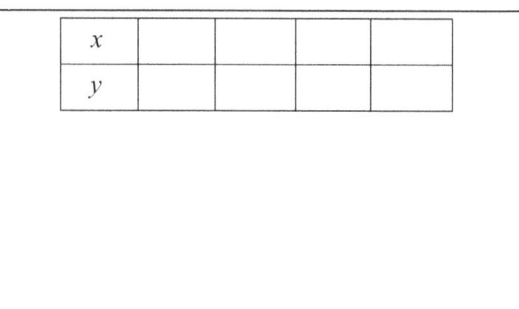

4)

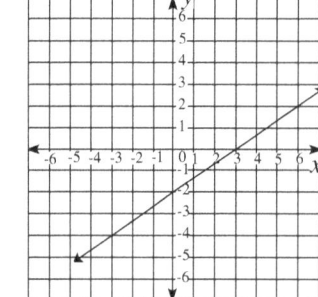

x				
y				

* For Questions **1-2**, use the graph below.

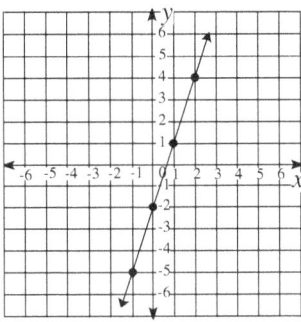

1. The graph shows the coordinates of $(0, -2)$ and $(1, 1)$. Which of the following coordinates are on the same line?

   A. $(2, 2)$ **B.** $(2, 4)$
   C. $(3, 4)$ **D.** $(3, 3)$

2. Which of the following represents the equation of the graph?

   A. $y = 2x + 1$ **B.** $y = 2x - 1$
   C. $y = 3x - 1$ **D.** $y = 3x - 2$

3. There is a linear function containing the coordinates of $(-3, -1)$, $(-1, 1)$, and $(1, 3)$. What is the equation of the graph?

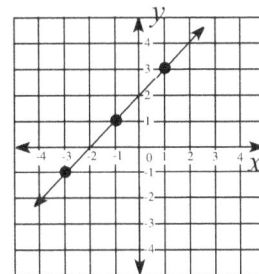

   A. $y = 2x + 2$ **B.** $y = x + 2$
   C. $y = 2x$ **D.** $y = 4x - 2$

4. Which of the following represents the equation $y = \dfrac{1}{2}x - 1$?

A.

x	0	2	4
y	−1	0	1

B.

x	−2	0	8
y	−2	−1	4

C.

x	8	10	12
y	2	3	4

D.

x	−4	0	1
y	−2	0	2

5. Which of the following represents the equation $y = 2x - 1$?

A.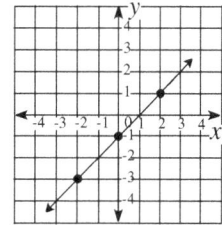

B.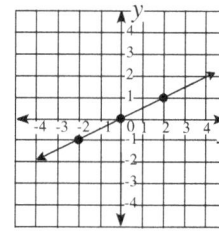

C.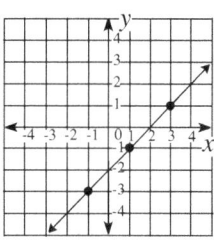

D.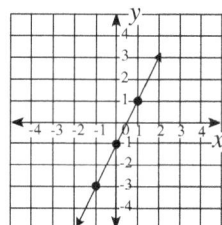

6. Which of the following represents the equation $y = \frac{1}{4}x + 3$?

A.

x	−20	−12	−8
y	−2	0	1

B.

x	32	40	48
y	8	10	12

C.

x	4	8	12
y	4	5	7

D.

x	−16	−10	−8
y	−1	0	1

7. Which of the following represents the equation $y = x + 2$?

A.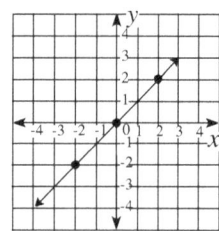

B.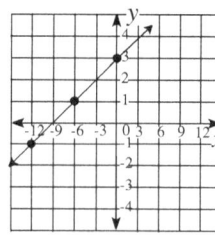

C.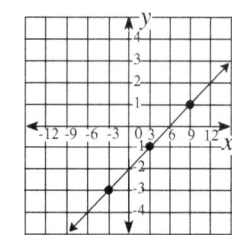

D.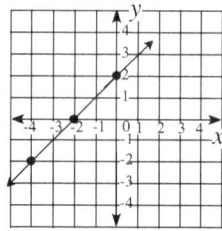

8. Which of the following graphs shows an increase in the values?

A.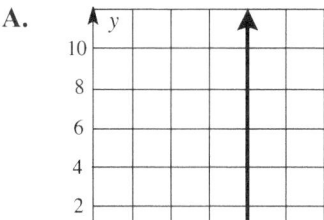

B.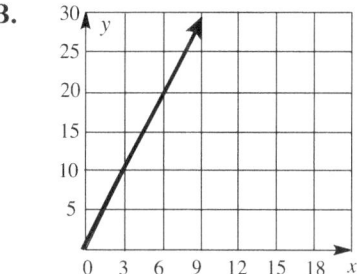

C.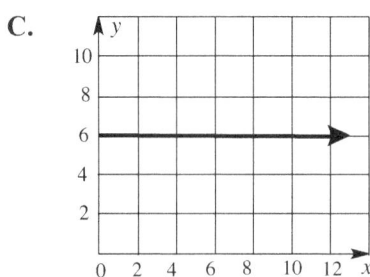

D.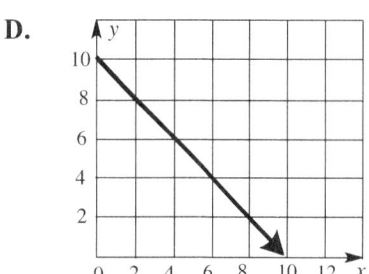

9. What is the equation of the graph below?

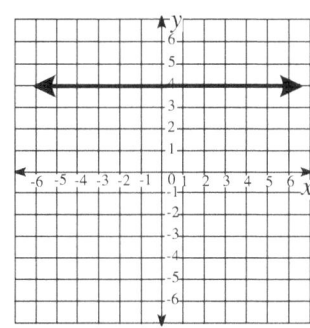

A. $y = x + 4$ B. $y = 4$

C. $y = 4x$ D. $x = 3$

10. What is the equation of the graph below?

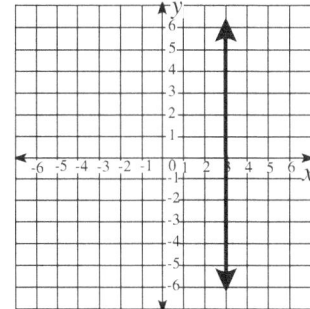

A. $y = x + 3$ B. Undefined.

C. $y = 3x$ D. $x = 3$

2. Slope, *x*- and *y*-intercepts

5–4. How to find the slope and *y*-intercept.

i) How to find the slope given a linear equation.
ii) How to find the *x*-intercept and *y*-intercept given a linear equation.
iii) How to find the slope given a graph.

---

i) Finding the slope of any linear equation.
  Any linear equation can be in the form $y = mx + b$, where <u>*m* is the slope</u> and <u>*b* is the *y*-intercept</u>.

ii) Finding the *x*-intercept and *y*-intercept given a linear equation.
  The *x*-intercept is where the graph crosses the *x*-axis and the *y*-intercept is where the graph crosses the *y*-axis.

The form of the linear equation is:
  $y = mx + b$, when $x = 0$, then the *y*-value is the *y*-intercept.
  $y = mx + b$, when $y = 0$, then the *x*-value is the *x*-intercept.

To find the *x*- and *y*-intercept of a linear equation on a graph:

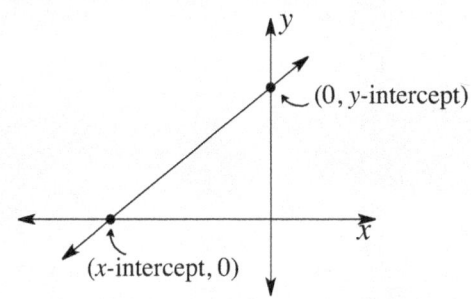

5–5. Find the slope, *x*-intercept, and *y*-intercept of the line.

$$y = 4x + 3$$

SOLUTION

a) First, look at the equation of $y = 4x + 3$.
  The slope-intercept form of a linear equation, $y = mx + b$, where *m* is the slope and *b* is the *y*-intercept.
So the slope of the line is $m = 4$.

b) To find the *x*- and *y*-intercepts.
i) When $x = 0$,
  $\qquad y = 4x + 3$ Original equation
  $\qquad y = 4(0) + 3$ Replace *x* with 0.
  $\qquad y = 3$

So the y-intercept is 3.

ii) When $y = 0$,

$$y = 4x + 3 \qquad \text{Original equation}$$
$$0 = 4x + 3 \qquad \text{Replace } y \text{ with 0.}$$
$$-3 = 4x + 3 - 3 \qquad \text{Subtract 3 from both sides.}$$
$$-3 = 4x \qquad \text{Simplify.}$$
$$-\frac{3}{4} = x \qquad \text{Divide each side by 4.}$$

So x-intercept is $-\frac{3}{4}$.

Quick Exercises 4   Find the slope, x-intercept, and the y-intercept of each equation.

1) $y = -x + 3$

2) $y = \frac{1}{2}x + 1$

3) $y = 2x - 3$

4) $y = -\frac{3}{4}x$

5-6. How to find the slope given a graph.

To find the slope of a nonvertical line.

$Slope\ (m) = \dfrac{\text{rise}}{\text{run}}$

$m = \dfrac{y_2 - y_1}{x_2 - x_1}$

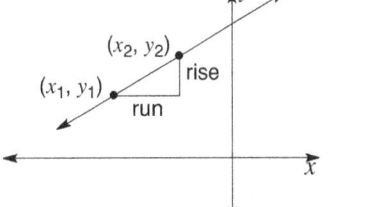

5-7. a) Find the slope of $\overleftrightarrow{AB}$ below. b) Find the x and y-intercepts of $\overleftrightarrow{AB}$ below. b) Find the linear equation given the information from a) and b).

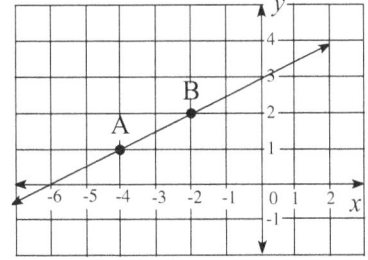

SOLUTION

a) Find the slope of AB.

As the line passes through A(−4, 1) and B(−2, 2), then the slope is given as:

Let (x_1, y_1) be A(−4, 1) and (x_2, y_2) be B(−2, 2).

$$m_{AB} = \frac{y_2 - y_1}{x_2 - x_1}$$ Apply the slope formula.

$$m_{AB} = \frac{2 - 1}{(-2) - (-4)} = \frac{1}{2}$$ Substitute in the values of the coordinates.

So, the slope of line m is $\frac{1}{2}$.

b) To find the x-intercept of AB.

The x-intercept is where the graph crosses the x-axis and the y-intercept is where the graph crosses the y-axis. When $x = 0$, then $y = 3$. So the y-intercept is 3. When $y = 0$, then $x = -6$. So the x-intercept is −6.

c) The linear equation in slope-intercept form is $y = mx + b$, where m is the slope and b is the y-intercept. So, the equation of line AB is $y = \frac{1}{2}x + 3$.

Quick Exercises 5   a) Find the slope of the line $\overleftrightarrow{AB}$ below. b) Find the x- and y-intercepts of $\overleftrightarrow{AB}$ below. b) Find the linear equation given the information from a) and b).

1)

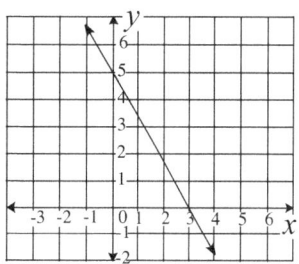

2)

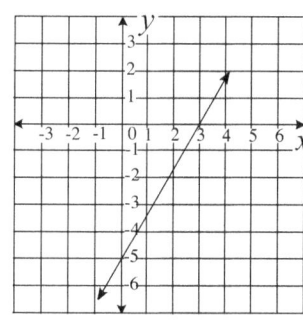

Slope:_____

x-intercept: _____

y-intercept: _____

Slope:_____

x-intercept: _____

y-intercept: _____

5-8. Write the linear equation using the given information below.

$$m = -3, \text{ }y\text{-intercept} = -1$$

SOLUTION

The linear equation in the form is $y = mx + b$, where m is the slope and b is the y-intercept. So, the equation is $y = -3x - 1$.

Quick Exercises 6   Write a linear equation using the given information.

1) $m = 2, y\text{-intercept} = 2$

2) $m = 3, y\text{-intercept} = -\dfrac{1}{2}$

3) $m = -\dfrac{1}{3}, y\text{-intercept} = 1$

4) $m = -2, y\text{-intercept} = -1$

5-9. Determine the slope of the line that passes through the points below.

$$A(3, 4) \text{ and } B(-5, -2)$$

SOLUTION

The line passes through points $A(x_1, y_1)$ and $B(x_2, y_2)$ as points $A(3, 4)$ and $B(-5, -2)$, Then the slope is given by:

$$m_{AB} = \frac{y_2 - y_1}{x_2 - x_1} \qquad \text{Apply the slope formula.}$$

$$m_{AB} = \frac{(-2) - 4}{(-5) - 3} = \frac{-6}{-8} \qquad \text{Substitute in the values of the coordinates.}$$

$$m_{AB} = \frac{3}{4} \qquad \text{Simplify.}$$

So, the slope of the line is $\dfrac{3}{4}$.

Quick Exercises 7   Find the slope of the line that passes through the points.

1) $A(-4, 5)$ and $B(0, -5)$

2) $R(-3, -4)$ and $S(-8, 3)$

3) $A(3, -5)$ and $B(4, 5)$

4) $R(5, 3)$ and $S(-2, -2)$

Exercises 5 Find the x and y-intercepts of each linear equation.

1) $y = x - 1$

x-intercept:

y-intercept:

2) $y = x + 2$

x-intercept:

y-intercept:

3) $y = x + 4$

x-intercept:

y-intercept:

4) $y = 2x - 3$

x-intercept:

y-intercept:

5) $y = -3x + 2$

x-intercept:

y-intercept:

6) $2y + 5 = -x$

x-intercept:

y-intercept:

7) $y - \frac{1}{2}x = 0$

x-intercept:

y-intercept:

8) $x - y = -3$

x-intercept:

y-intercept:

9) $x = -1$

x-intercept:

y-intercept:

10) $2y = -\frac{1}{2}x + 4$

x-intercept:

y-intercept:

11) $y = \frac{2}{5}x - 3$

x-intercept:

y-intercept:

12) $-y + 5x = \frac{1}{2}$

x-intercept:

y-intercept:

Exercises 6 Find the slope and y-intercepts of each linear equation.

1) $y = 2x + 1$

slope:

y-intercept:

2) $y = -2x - 1$

slope:

y-intercept:

3) $y = x + 4$

slope:

y-intercept:

4) $y + x = 3$

slope:

y-intercept:

5) $y = -2x - 3$

slope:

y-intercept:

6) $-y + 2 = -5x$

slope:

y-intercept:

7) $2x + y = 1$

slope:

y-intercept:

8) $y - 3x + 2 = 0$

slope:

y-intercept:

9) $y = -1$

slope:

y-intercept:

10) $y = -\dfrac{1}{2}x - 3$

slope:

y-intercept:

11) $\dfrac{1}{3}x - 3y = -2$

slope:

y-intercept:

12) $-5x - y = \dfrac{1}{2}$

slope:

y-intercept:

Exercises 7 Find the slope, *x* and *y*-intercepts of each graph.

1)

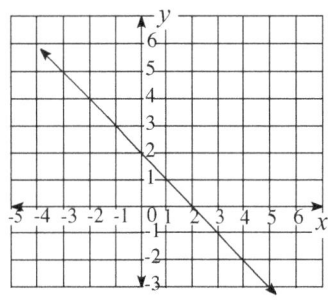

Slope:_____

x-intercept: _____

y-intercept: _____

2)

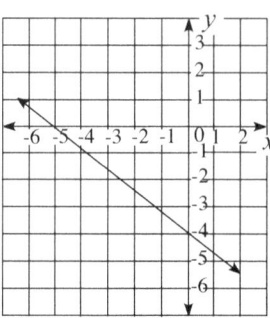

Slope:_____

x-intercept: _____

y-intercept: _____

3)

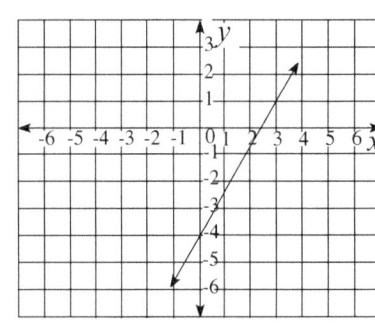

Slope:_____

x-intercept: _____

y-intercept: _____

4)

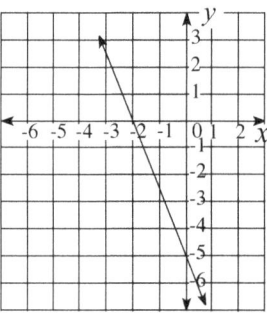

Slope:_____

x-intercept: _____

y-intercept: _____

5)

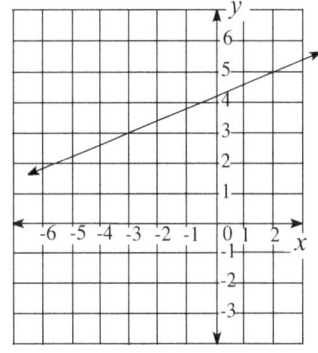

6)

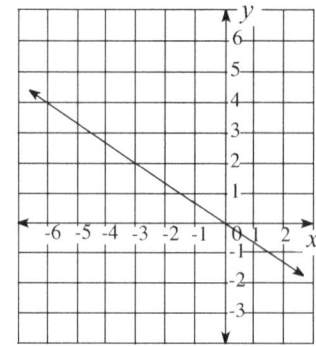

Slope:_____   Slope:_____

x-intercept: _____ x-intercept: _____

y-intercept: _____ y-intercept: _____

Exercises 8   Find the linear equation of each graph.

1)
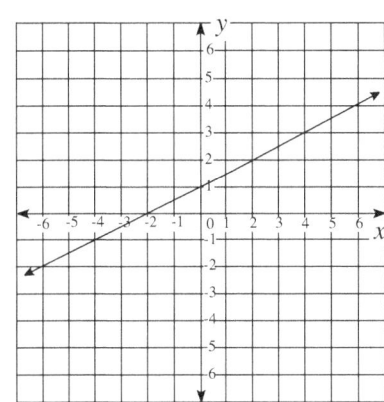

2)
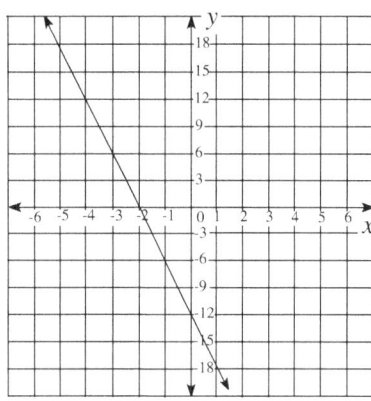

3)
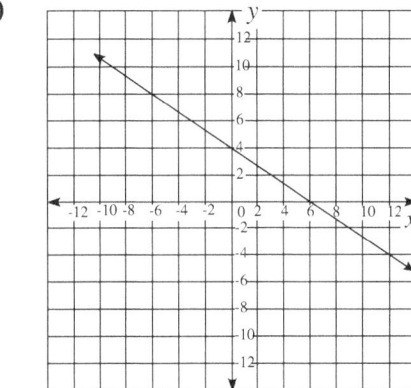

4)
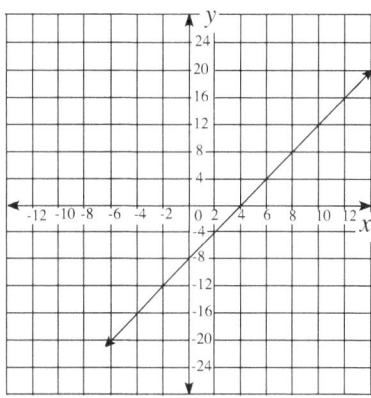

5)
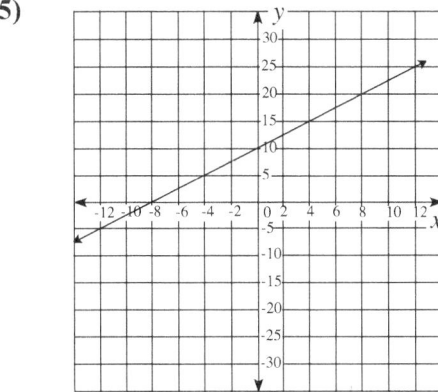

6)
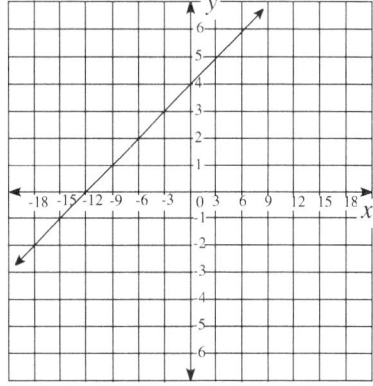

Exercises 9 Find the linear equation of each graph. Explain how you got your answer.

1)
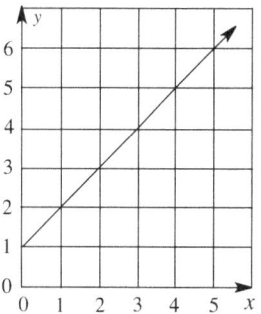

2)
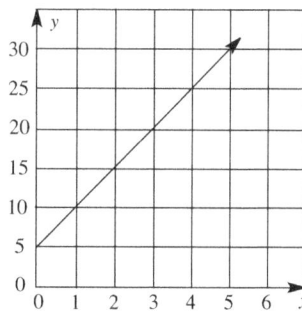

3)
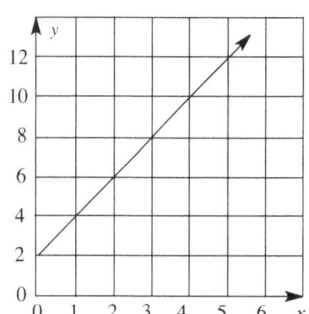

4)
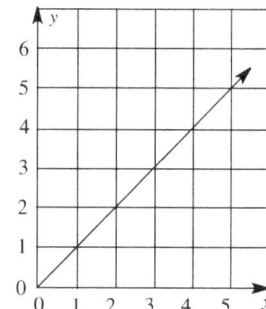

5)
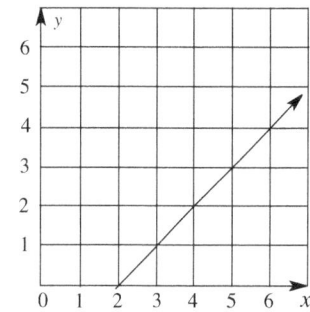

6)
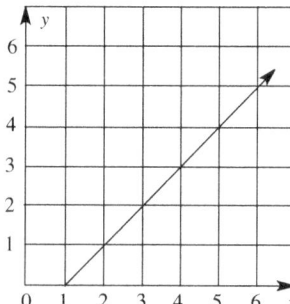

7)
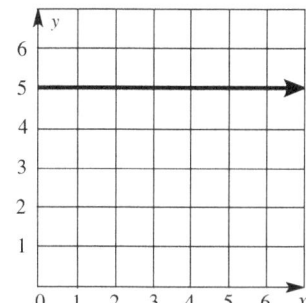

8)
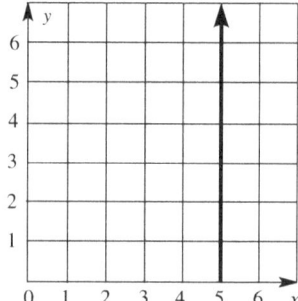

Exercises 10 Make a linear equation using the given information.

1) $m = 4$, y-intercept $= 1$

2) $m = -1$, y-intercept $= -2$

3) $m = \dfrac{1}{3}$, y-intercept $= -\dfrac{2}{3}$

4) $m = -\dfrac{1}{2}$, y-intercept $= 7$

5) $m = -3$, y-intercept $= 4$

6) $m = 1$, y-intercept $= -3$

Exercises 11 Find the slope of the line that passes through the given points.

1) $(1, 2)$ and $(-3, 4)$

2) $(-3, 3)$ and $(1, -2)$

3) $(5, 3)$ and $(-3, 0)$

4) $(2, 3)$ and $(-3, 2)$

5) $(2, 7)$ and $(-1, -3)$

6) $(-1, 4)$ and $(4, -5)$

7) $(3, 1)$ and $(-5, -9)$

8) $(-8, 2)$ and $(0, -12)$

Exercises 12 Find the slope of each graph.

1)
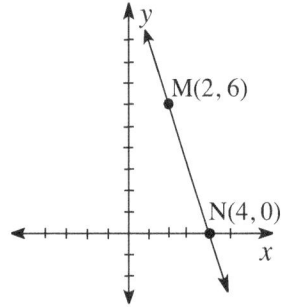

2)
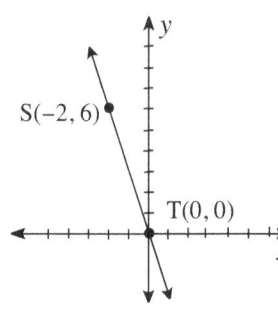

3)
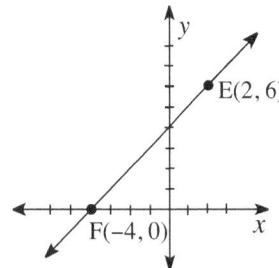

4)
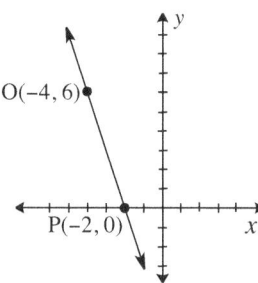

5)
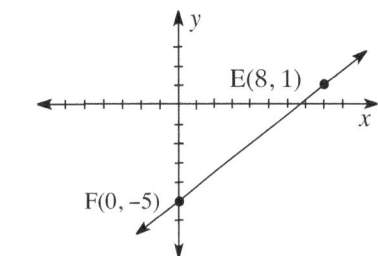

6)
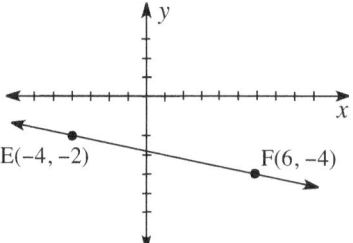

Exercises 13 Find the x-intercept and y-intercept of the line that passes through the given points.

1) $(-1, 3)$ and $(-3, 1)$

2) $(1, 4)$ and $(-2, 2)$

_____

_____

3) $(0, 3)$ and $(5, -2)$

4) $(-2, -2)$ and $(0, -4)$

_____

_____

1. What is the slope of $y = 2x - 8$?

   A. 8   B. −8
   C. 2   D. −2

2. What is the y-intercept of $y = \frac{1}{2}x + 3$?

   A. $\frac{1}{2}$ B. $-\frac{1}{2}$
   C. 3   D. −3

3. What is the slope of $2x - 3y = 1$?

   A. 2   B. −3
   C. $\frac{2}{3}$ D. $-\frac{2}{3}$

4. What is the slope of $-6x + 2y = 4$?

   A. 3   B. −3
   C. $\frac{1}{3}$ D. $-\frac{1}{3}$

5. What are the x- and y-intercepts of $y = 2x - 1$?

   A. x-intercepts = 2, y-intercepts = −1   B. x-intercepts = $-\frac{1}{2}$, y-intercepts = −1

   C. x-intercepts = $\frac{1}{2}$, y-intercepts = −1   D. x-intercepts = 2, y-intercepts = $-\frac{1}{2}$

6. Which of the following are the x- and y-intercepts of $-2x + y = 1$?

   A. $(\frac{1}{2}, 0), (0, 1)$ B. $(-\frac{1}{2}, 0), (0, -1)$
   C. $(-2, 0), (0, 1)$ D. $(1, 2), (0, -1)$

7. What are the slope and y-intercept of $y - \frac{2}{3}x + 4 = 1$?

   A. $m = \frac{2}{3}$, y-intercepts = 1   B. $m = 1$, y-intercepts = 1
   C. $m = \frac{2}{3}$, y-intercepts = −3   D. $m = -\frac{2}{3}$, y-intercepts = −3

8. Which of the following is the equation of a line with a slope of −3 and a *y*-intercept of 2?

    A. $y = -3x + 2$ **B.** $y = 2x - 3$

    C. $y = 3x + 2$ **D.** $y = -2x + 3$

9. Which of the following is the equation of a line with a slope of $\frac{2}{3}$ and a *y*-intercept of 2?

    A. $y = -\frac{2}{3}x + 2$ **B.** $y = 2x + \frac{2}{3}$

    C. $y = \frac{2}{3}x + 2$ **D.** $y = 2x - \frac{2}{3}$

10. What is the slope of the line in the graph below?

    A. −2   **B.** $-\frac{1}{2}$

    C. 1   **D.** $-1\frac{1}{2}$

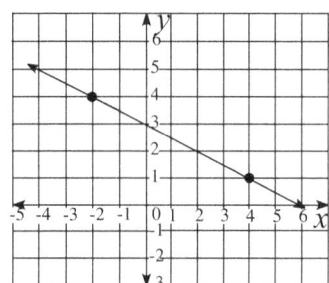

11. What is the slope of the line in the graph below?

    A. −2   **B.** $-\frac{1}{2}$

    C. 1   **D.** $-2\frac{1}{2}$

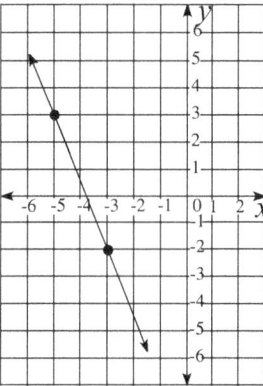

12. Which of the following is the slope of a line that passes through the given points (2, 3) and (−3, 2)?

    A. $-\frac{1}{5}$ **B.** $\frac{1}{5}$

    C. $-\frac{1}{4}$ **D.** $\frac{1}{4}$

13. Which of the following is the slope of a line that passes through the given points (2, 3) and (−3, 2)?

    A. $-\dfrac{1}{6}$ **B.** $\dfrac{1}{6}$

    C. $-\dfrac{1}{4}$ **D.** 1

14. Which of the following is the *y*-intercept of a line that passes through the given points (−2, 3) and (0, 2)?

    A. $-\dfrac{1}{5}$ **B.** $\dfrac{1}{5}$

    C. $-\dfrac{1}{2}$ **D.** $\dfrac{1}{4}$

15. Which of the following is the *x*-intercept of a line that passes through the given points (−5, 0) and (1, 9)?

    A. −3   **B.** 3

    C. $-\dfrac{1}{3}$ **D.** $\dfrac{1}{3}$

16. Which of the following is the value of *y* of a line that has a given slope of −2 and passes through the given points (−1, *y*) and (1, 0)?

    A. −4   **B.** $\dfrac{1}{4}$

    C. 2   **D.** 4

17. Which of the following is the value of *x* of a line that has a given slope of $\dfrac{2}{3}$ and passes through the given points (−4, −1) and (*x*, 3)?

    A. 2   **B.** $\dfrac{2}{3}$

    C. 1   **D.** −1

18. Which of the following is the equation of a line that passes through P(3, −2) and has a slope of 1?

    A. $y = x - 5$ **B.** $y = x + 1$

    C. $y = -x - 5$ **D.** $y = x - 1$

19. Which of the following is the equation of a line that passes through P(−4, −4) and has a slope of $\frac{1}{4}$?

A. $y = \frac{1}{4}x - 3$ **B.** $y = \frac{1}{4}x + 3$

C. $y = -\frac{1}{4}x - 4$ **D.** $y = -4x - 3$

20. What is the slope of the line in the graph below?

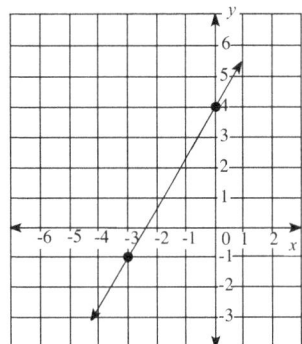

A. −2   **B.** $-\frac{1}{2}$

C. 1   **D.** $-1\frac{1}{2}$

21. What is the *y*-intercept of the line in the graph below?

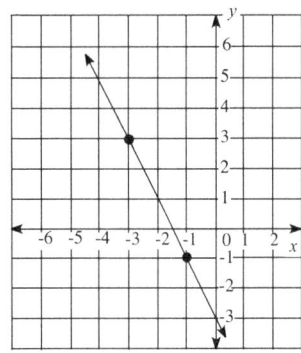

A. −2   **B.** $-\frac{1}{2}$

C. 1   **D.** $-1\frac{1}{2}$

22. Which of the following graphs represents $y = -x - 7$?

A) 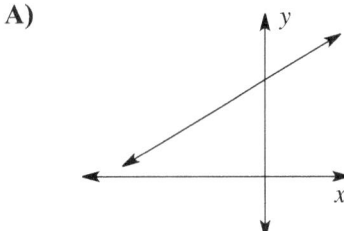   **B)**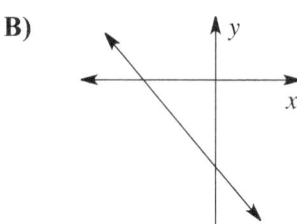

C)

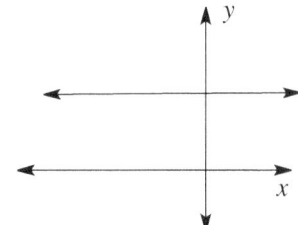

D)

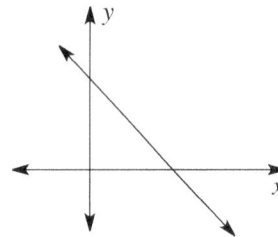

23. Which of the following graphs represents $y = -x + 5$?

A)

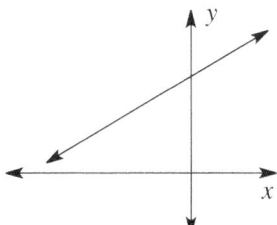

B)

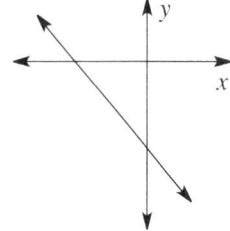

C)

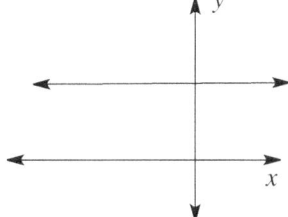

D)

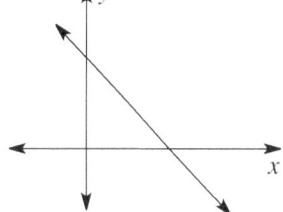

3. Parallel Lines

5-10. Determine whether if the two lines are parallel.

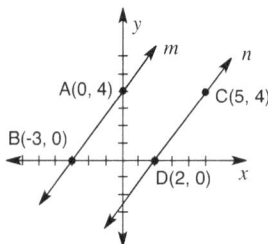

SOLUTION

Let's find the slopes of the two lines.
Line m passes through A(0, 4) and B(−3, 0).

$$m_{AB} = \frac{y_2 - y_1}{x_2 - x_1}$$ Apply the slope formula.

$$m_{AB} = \frac{0 - 4}{(-3) - 0} = \frac{4}{3}$$ Substitute in the values of the coordinates.

So, the slope of line m is $\frac{4}{3}$.

Line m passes through C(5, 4) and D(2, 0).

$$m_{CD} = \frac{y_2 - y_1}{x_2 - x_1}$$ Apply the slope formula.

$$m_{CD} = \frac{0 - 4}{2 - 5} = \frac{-4}{-3} = \frac{4}{3}$$ Substitute in the values of the coordinates.

So, the slope of line m is $\frac{4}{3}$.
Based on the results of two slopes, lines m and n are parallel because parallel lines always have congruent slopes.

Quick Exercises 8   Determine whether if the two lines are parallel.

1)

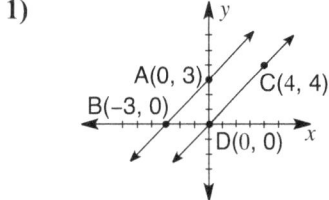

2)

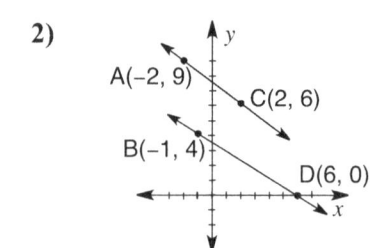

4. Perpendicular Lines

5-11. Find if ST is perpendicular to PQ.

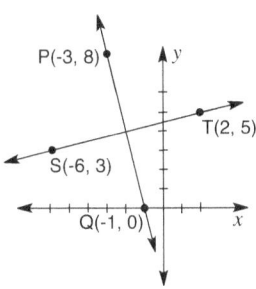

SOLUTION

In a coordinate plane, two lines are perpendicular if and only if **the product of their slopes is −1**. Vertical and horizontal lines are perpendicular.

(slope 1) × (slope 2) = **−1**

Find the slopes of the lines ST and PQ. You can check if the lines are perpendicular if (−1) is the product of their slopes.

The line passes through S(−6, 3) and T(2, 5).

$S(x_1, y_1)$ and $T(x_2, y_2)$. $m_1 = \overleftrightarrow{ST} = \dfrac{y_2 - y_1}{x_2 - x_1}$ Apply the slope formula.

$m_1 = \dfrac{5 - 3}{2 - (-6)} = \dfrac{2}{8} = \dfrac{1}{4}$ Substitute in the given information and calculate.

So, the slope of $\overleftrightarrow{ST}$ is $\dfrac{1}{4}$.

The line passes through P(−3, 8) and Q(−1, 0).

$m_2 = \overleftrightarrow{PQ} = \dfrac{y_2 - y_1}{x_2 - x_1}$ Apply the slope formula.

$m_2 = \dfrac{0 - 8}{-1 - (-3)} = \dfrac{-8}{2} = -4$ Substitute in the given values and solve.

So, that the slope of $\overleftrightarrow{PQ}$ is −4.

Find the product of m_1 and m_2.

$m_1 \times m_2 = \left(\dfrac{1}{4}\right) \times -4 = -1$, so $\overleftrightarrow{ST}$ and $\overleftrightarrow{PQ}$ are perpendicular.

Quick Exercises 9   Determine whether if the two lines are perpendicular.

1)

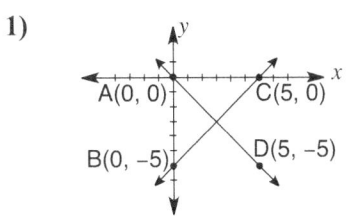

2)

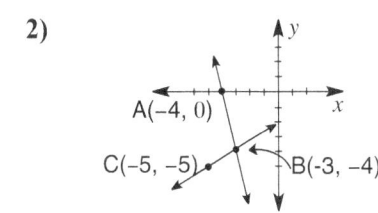

5–12. Find the equation of the line that passes through P and is perpendicular to the given equation.

$$P(5, -4), \ y = 2x + 3$$

SOLUTION

Suppose that the first line is v and is represented by the equation $y = 2x + 3$. The second line is w and the equation is unknown.

The product of the slope of the perpendicular lines is -1. Find each slope of v and w.

The slope of m_v is 2.

$$\text{So } (m_v)(m_w) = -1$$
$$(2)(m_w) = -1$$
$$m_w = -\frac{1}{2}$$

Now use the slope $m_w = -\frac{1}{2}$ to find the y-intercept.

$(x, y) = P(5, -4)$	Given
$y = mx + b$	Equation of a line
$-4 = (-\frac{1}{2})(5) + b$	Substitute in the values.
$-4 = (-\frac{5}{2}) + b$	Simplify.
$-\frac{3}{2} = b$	Add $-\frac{5}{2}$ to both sides.

So the equation of line w is $y = -\frac{1}{2}x - \frac{3}{2}$.

Quick Exercises 10   a) Find the equation of the line that passes through P and is perpendicular to the given equation.

1) $P(1, 1), y = -2x + 1$

2) $P(-1, 1), y = -x + 2$

3) $P(0, 1), y = x + 2$

4) $P(-1, 0), y = 2x + 1$

b) Determine whether if the two given lines are perpendicular.

5) line m: $y = x - 2$
line n: $-2y = x - 1$

6) line m: $2x = y - 3$
line n: $2y = -x + 9$

Exercises 14 Determine whether if the two given lines are parallel.

1)

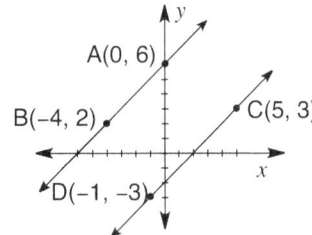

2)

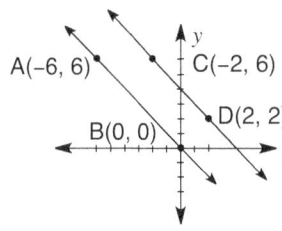

3)

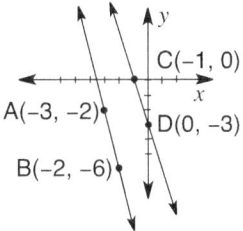

4)

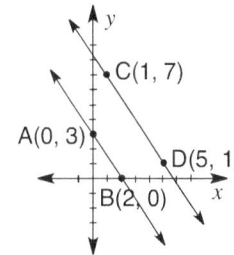

5)

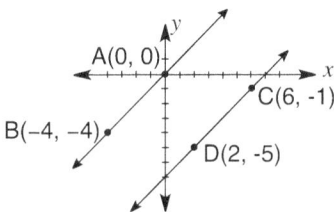

6)

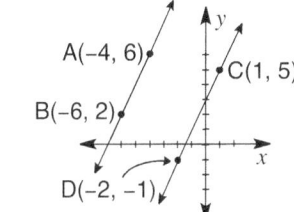

Exercises 15 Find each linear equation given the slope and the coordinates of the point that the line passes through.

1) $P(-3, 2)$, $m = 0$

2) $P(-5, -4)$, $m = -3$

3) $P(2, 5)$, $m = -3$

4) $P(-2, 2)$, $m = 1$

5) $P(-1, 2)$, $m = 1$

6) $P(\frac{3}{4}, -1)$, $m = -\frac{1}{2}$

7) $P(-\frac{1}{3}, 2)$, $m = 3$

8) $P(-2, -4)$, $m = -1$

Exercises 16 Given a line with a y-intercept of 7 and is parallel to each of the lines below, find its equation.

1) $y + 3x = 4$

2) $-x + 5y = 8$

3) $5x - 2y = -3$

4) $-3y - 6x = 3$

5) $y = x + 4$

6) $-x + y = 2$

7) $3x + 2y = 1$

8) $3y - x = 0$

Exercises 17 Find the equation of the line that passes through P and is perpendicular to the given equation.

1) $P(0, 0), y = -4x + 2$

2) $P(-1, -6), y = 3x - 1$

3) $P(-2, -1), y = -x - 4$

4) $P(3, -3), y = -4x - 1$

5) $P(-1, 2), y = 3x - 5$

6) $P(\frac{3}{4}, -1), y = 2x - 3$

7) $P(-\frac{1}{3}, 2), y = 5x - 2$

8) $P(-2, -4), y = -3x + 1$

Exercises 18 Determine whether if the two lines are perpendicular.

1)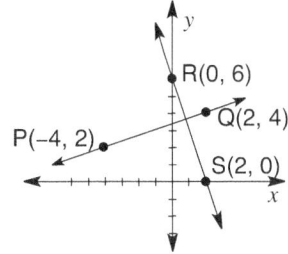

2)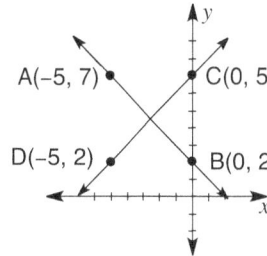

3)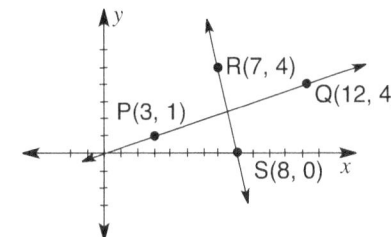

4)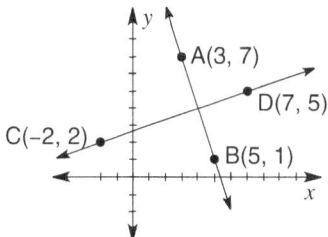

Exercises 19 Determine whether if the two given lines are perpendicular.

1) line m: $4x = y - 1$
   line n: $3x = 6y - 5$

2) line m: $2x = 2y - 1$
   line n: $3x = 6y - 5$

3) line m: $-3x = y - 2$
   line n: $3x = 9y - 4$

4) line m: $3x = -2y - 4$
   line n: $-x = 4y - 3$

5) line m: $3x = 6y - 5$
   line n: $-10x = 5y - 4$

6) line m: $2x = 2y - 1$
   line n: $2y = -6x + 8$

7) line m: $x = 2y - 2$
   line n: $-2x = y - 1$

8) line m: $5x = 10y - 3$
   line n: $10y = -5x + 9$

Exercises 20 Determine if the slopes are perpendicular.

1) $m_1 = -5$

   $m_2 = \dfrac{1}{5}$

2) $m_1 = -\dfrac{2}{3}$

   $m_2 = \dfrac{3}{2}$

3) $m_1 = -3$

   $m_2 = \dfrac{1}{3}$

4) $m_1 = -2$

   $m_2 = -\dfrac{1}{2}$

5) $m_1 = -1$

   $m_2 = 0$

6) $m_1 = -\dfrac{5}{8}$

   $m_2 = \dfrac{5}{8}$

7) $m_1 = -3$

   $m_2 = \dfrac{3}{2}$

8) $m_1 = -4$

   $m_2 = \dfrac{1}{4}$

Exercises 21 Determine if the given lines are parallel or perpendicular.

1) $y = -x + 2$

   $\dfrac{1}{2}x + 2y + 3 = 0$

2) $y = 2x - 1$

   $4x - 2y + 3 = 0$

3) $y = -x - 4$

   $\dfrac{3}{2}x - 3y - 2 = 0$

4) $y = 2x - 1$

   $5y = -\dfrac{5}{2}x + 2$

5) $y = 3x - 5$

   $x + 3y + 3 = 0$

6) $y = 2x - 3$

   $2x - y + 1 = 0$

7) $y = -x - 2$

   $3x - 3y + 1 = 0$

8) $y = -3x + 1$

   $3x - 9y + 1 = 0$

1. Which of the following is the equation of a line that passes through $P(\frac{1}{2}, -1)$ and has a slope of $\frac{1}{2}$?

    A. $y = -\frac{1}{2}x - \frac{5}{4}$ **B.** $y = -\frac{1}{2}x - 1$

    C. $y = -x + \frac{1}{2}$ **D.** $y = \frac{1}{2}x - \frac{5}{4}$

2. Use the graph to find the slopes of the lines and determine if $\overleftrightarrow{AB}$ is perpendicular to $\overleftrightarrow{DC}$.

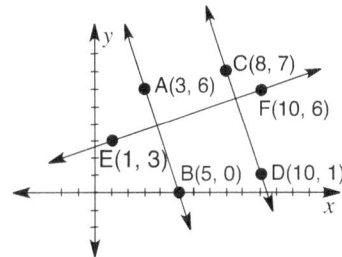

    A. $\overleftrightarrow{AB} = -\frac{2}{3}$, $\overleftrightarrow{CD} = -\frac{2}{3}$, $\overleftrightarrow{EF} = \frac{3}{2}$, yes   **B.** $\overleftrightarrow{AB} = -3$, $\overleftrightarrow{CD} = -3$, $\overleftrightarrow{EF} = \frac{1}{3}$, yes

    C. $\overleftrightarrow{AB} = -\frac{1}{3}$, $\overleftrightarrow{CD} = -\frac{1}{3}$, $\overleftrightarrow{EF} = 3$, yes   **D.** $\overleftrightarrow{AB} = \frac{1}{3}$, $\overleftrightarrow{CD} = \frac{1}{3}$, $\overleftrightarrow{EF} = 3$, no

3. If the y-intercept of a given line is 7 and is parallel to the equation $5x - 2y = -3$, which of the following is the equation of the line?

    A. $y = \frac{5}{6}x + 7$ **B.** $5y - 2x = 14$

    C. $2y - 5x = 14$ **D.** $y = \frac{2}{3}x - 7$

4. If the y-intercept of a given line is $\frac{3}{4}$ and is parallel to the equation $-3y - 6x = 3$, which of the following is the equation of the line?

    A. $8x - 4y = 3$ **B.** $y = -2x + \frac{3}{4}$

    C. $6x + 2y = 4$ **D.** $y = 2x + \frac{3}{4}$

5. Use the graph to find the slopes of the lines.

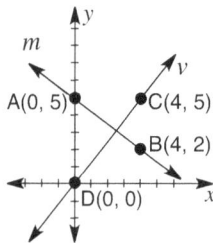

A. $\overleftrightarrow{AB} = -\dfrac{3}{4}$, $\overleftrightarrow{CD} = 1$ **B.** $\overleftrightarrow{AB} = -\dfrac{3}{4}$, $\overleftrightarrow{CD} = \dfrac{3}{4}$

C. $\overleftrightarrow{AB} = -1$, $\overleftrightarrow{CD} = 1$ **D.** $\overleftrightarrow{AB} = -\dfrac{3}{4}$, $\overleftrightarrow{CD} = \dfrac{5}{4}$

6. Which of the following best describes the slope of perpendicular lines?

    A. Two lines are perpendicular if and only if the product of their slopes is -1.
    B. Two lines are perpendicular if and only if the product of their slopes is 0.
    C. Two lines are perpendicular if and only if the product of their slopes is $-\dfrac{1}{2}$.
    D. Two lines are perpendicular if and only if the product of their slopes is 1.

7. Which of the following is the slope of a line perpendicular to line p with a slope of -2?

    A. -2 **B.** 2
    C. $-\dfrac{1}{2}$ **D.** $\dfrac{1}{2}$

8. Use the graph to find the slopes of the lines and determine if $\overleftrightarrow{AB}$ is perpendicular to $\overleftrightarrow{DC}$.

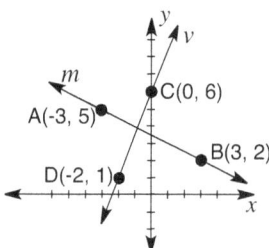

A. $\overleftrightarrow{AB} = \dfrac{5}{4}$, $\overleftrightarrow{CD} = \dfrac{1}{2}$, no   **B.** $\overleftrightarrow{AB} = -\dfrac{1}{2}$, $\overleftrightarrow{CD} = \dfrac{5}{2}$, no

C. $\overleftrightarrow{AB} = \dfrac{5}{2}$, $\overleftrightarrow{CD} = -\dfrac{1}{4}$, no   **D.** $\overleftrightarrow{AB} = -\dfrac{1}{2}$, $\overleftrightarrow{CD} = 2$, yes

9. If the *y*-intercept of a given line is –4 and is parallel to the equation $-x + 5y = 8$, which of the following is the equation of the line?

    A. $5y - 8 = x$ **B.** $10y - 2x = 4$

    C. $y = 5x - 3$ **D.** $5y = x - 20$

10. Which of the following is the slope of a line perpendicular to line *p* with a slope of $-\frac{1}{4}$?

    A. -4 **B.** 4

    C. $-\frac{1}{4}$ **D.** $\frac{1}{4}$

11. Use the graph to find the slopes of the lines.

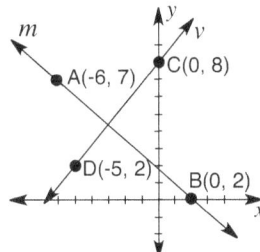

    A. $\overleftrightarrow{AB} = -\frac{5}{6}, \ \overleftrightarrow{CD} = 1$ **B.** $\overleftrightarrow{AB} = -\frac{6}{5}, \ \overleftrightarrow{CD} = \frac{6}{5}$

    C. $\overleftrightarrow{AB} = -1, \ \overleftrightarrow{CD} = 1$ **D.** $\overleftrightarrow{AB} = -\frac{5}{6}, \ \overleftrightarrow{CD} = \frac{6}{5}$

12. Use the graph to find the slopes of the lines.

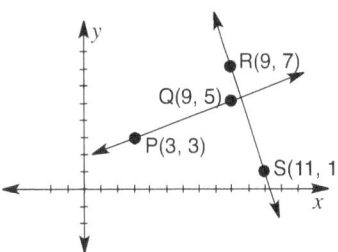

    A. $\overleftrightarrow{PQ} = -\frac{1}{3}, \ \overleftrightarrow{RS} = 3$ **B.** $\overleftrightarrow{PQ} = 1, \ \overleftrightarrow{RS} = -1$

    C. $\overleftrightarrow{PQ} = \frac{1}{3}, \ \overleftrightarrow{RS} = -3$ **D.** $\overleftrightarrow{PQ} = -1, \ \overleftrightarrow{RS} = 1$

ANSWERS

CHAPTER 1

Quick Exercises 1
1) Natural Number   2) Irrational Number   3) Whole Number   4) Rational Number

Quick Exercises 2
1) y cubed times x to the fourth power   2) 2 cubed times x squared   3) x^7 $(4)(2^4)$

Quick Exercises 3
1) 5   2) 3   3) 16   4) 4

Quick Exercises 4
1) 29.1   2) 47   3) 11   4) 4

Quick Exercises 5
1) 6   2) 16   3) 5   4) 1

Exercises 1
1) 2 to the fourth power   2) x to the fifth power   3) 5 times 2 cubed
4) 3 times a squared   5) 4 times 2 to the fourth power   6) $4y^6$
7) $(2)(2^2)$ 8) two-third squared   9) $2xy^2$
10) m cubed times n to the fourth power   11) 1 for any number x $(x \neq 0)$ 12) 2

Exercises 2
1) 4   2) 32   3) 4   4) 8   5) 11   6) 8
7) 20   8) 5   9) 6   10) 6   11) 11   12) 17

Exercises 3
1) -40   2) Irrational Number   3) Irrational Number
4) rational Number   5) Irrational Number   6) rational Number
7) whole Number   8) Real Number   9) Natural Number
10) Natural Number

Exercises 4
1) 1   2) 3   3) 14   4) 5   5) 9   6) 8
7) -10   8) 0   9) 2.5   10) 15

Exercises 5
1) 4   2) -12   3) -10   4) 1.5   5) 3   6) -5
7) 4   8) 2   9) 12   10) -1   11) 2   12) -4.5

Exercises 6
1) 10   2) 15   3) 20   4) 1   5) $(\frac{1}{2})$ 6) $-8\frac{1}{2}$

CHAPTER 2

Quick Exercises 1
1)   -5   2)   -1   3)   1   4)   5

Quick Exercises 2
1)   5   2)   -5   3)   -1   4)   1

Quick Exercises 3
1)   54   2)   -8   3)   40   4)   90

Quick Exercises 4
1)   -2   2)   -1   3)   -13   4)   13

Quick Exercises 5
1)   2.54   2)   83.70   3)   -12.23   4)   -94.43

Quick Exercises 6
1)   1.11   2)   -7.6   3)   71.4   4)   1.53

Quick Exercises 7
1)   1   2)   -1   3)   2   4)   0

Quick Exercises 8
1) $1\frac{2}{3}$ 2)   2.75   3) $-\frac{1}{4}$ 4) $-3\frac{2}{5}$

Quick Exercises 9
1) $-1\frac{1}{2}$ 2) $+\frac{1}{3}$ 3) $+\frac{1}{8}$ 4) $2\frac{3}{5}$

Quick Exercises 10
1)   -10.8   2)   19.2   3)   -16   4)   13.5

Quick Exercises 11
1)   -0.4   2)   4.1   3)   -1.8   4)   -7.2

Quick Exercises 12
1)   -18   2)   -7

Quick Exercises 13
1)   2   2) $+\frac{1}{6}$

Quick Exercises 14
1) $-2\sqrt{7}$ 2)   -3   3) ± 7 4) $\left(\frac{2}{5}\right)$

Quick Exercises 15
1)   Distributive Property   2)   Associative Property of Addition   3)   Substitution   4)   Commutative Property of Addition

Quick Exercises 16
1) $2(x-20)$ 2) $5(2+3x)$ 3) $11x-4$ 4) $21-11xy+2x$

Exercises 1
1)	-13	2)	-8	3)	-27	4)	-19	5)	-36	6)	-52
7)	-$103	8)	-$158	9)	$67	10)	-$768				

Exercises 2
1)	1	2)	-11	3)	42	4)	-34	5)	-44	6)	20
7)	66	8)	32	9)	-66	10)	3	11)	643	12)	-288
13)	678	14)	527								

Exercises 3
1)	-27	2)	-28	3)	-40	4)	30	5)	100	6)	-496
7)	240	8)	-9								

Exercises 4
1)	-4	2)	18	3)	-21.25	4)	4	5)	20	6)	$5.25
7)	$35	8)	-24	9)	-12	10)	60				

Exercises 5
1)	4	2)	7.5	3)	12	4)	-13	5)	-1	6)	60
7)	4	8)	5	9)	-23	10)	-180				

Exercises 6
1)	3.23	2)	$83.7	3)	1.23	4)	-25.57	5)	-0.747	6)	-6.014
7)	-52.43	8)	-1.203	9)	17.05	10)	37.42				

Exercises 7
1)	45.95	2)	31.27	3)	-5.296	4)	7.76	5)	-38.75	6)	28.35
7)	0.11	8)	-0.24	9)	-4.29	10)	-3.24	11)	4.85	12)	27.57

Exercises 8
1)	0	2)	12	3)	6	4)	3	5)	7	6)	0
7)	0	8)	13.6								

Exercises 9
1)	-10	2)	3.25	3)	-2	4)	-1	5)	-3	6)	12
7)	4	8)	3	9)	11	10)	4				

Exercises 10
1)	2	2)	4	3)	$-3\frac{1}{2}$	4)	-2	5)	10	6)	-3
7)	-2	8)	-4								

Exercises 11
1)	$2\frac{2}{5}$	2)	$(\frac{3}{4})$	3)	$-2\frac{1}{3}$	4)	$3\frac{5}{8}$	5)	$4\frac{7}{8}$	6)	$7\frac{3}{4}$
7)	$4\frac{2}{3}$	8)	1.25	9)	$-2\frac{5}{9}$	10)	$5\frac{5}{6}$				

Exercises 12
1)	$3\frac{1}{3}$	2)	$1\frac{1}{7}$	3)	$3\frac{1}{4}$	4)	$(\frac{1}{5})$	5)	$(\frac{1}{3})$	6)	$(\frac{1}{3})$
7)	$1\frac{1}{5}$	8)	$-5\frac{2}{3}$								

Exercises 13
1) 637	**2)** 10	**3)** 1.4	**4)** $78.3604	**5)** -434	**6)** -10.8
7) 6	**8)** 17.26	**9)** 1.75	**10)** 18.6	**11)** -3	**12)** 125
13) 14.4	**14)** 2.5				

Exercises 14
1) 1.4	**2)** -4.8	**3)** -12	**4)** 0.07	**5)** 0.6	**6)** -0.2
7) 25	**8)** 0.25	**9)** -6.8	**10)** 0.56	**11)** 0.02	**12)** 4
13) 4	**14)** 0.21				

Exercises 15
1) 0.13   **2)** $3\frac{3}{7}$ **3)** $\frac{2}{15}$ **4)** $(\frac{3}{4})$ **5)** $\frac{4}{49}$ **6)** $\frac{5}{18}$

7) $(\frac{1}{4})$ **8)** $(\frac{2}{5})$ **9)** $\frac{2}{7}$ **10)** $(\frac{1}{6})$ **11)** $8\frac{11}{24}$ **12)** 1

Exercises 16
1) $(\frac{1}{7})$ **2)** $(\frac{13}{24})$ **3)** $(\frac{3}{56})$ **4)** 2   **5)** $(\frac{3}{4})$ **6)** $(\frac{1}{3})$

7) 3   **8)** $(\frac{49}{90})$ **9)** $(\frac{7}{10})$ **10)** $(\frac{3}{5})$ **11)** $(\frac{2}{7})$ **12)** $1\frac{19}{33}$

Exercises 17
1) $7\sqrt{2}$ **2)** $-3\sqrt{3}$ **3)** -6   **4)** $\pm 5\sqrt{2}$ **5)** $-4\sqrt{2}$ **6)** -0.7

7) $(\frac{2}{5})$ **8)** $\sqrt{2}$ **9)** $6\sqrt{3}$ **10)** $-2\sqrt{2}$ **11)** 10   **12)** $\pm 4\sqrt{2}$

Exercises 18
1) Distributive Property	**2)** Commutative Property Addition	**3)** Identity Property of Addition
4) Reflexive Property	**5)** Identity Property of Multiplication	**6)** Associate Property of Addition
7) Reflexive Property	**8)** Substitution	**9)** Substitution
10) Distributive Property	**11)** Reflexive	**12)** Identity Property of Multiplication
13) Reflexive Property	**14)** Distributive Property	

Exercises 19
1) $0 = 5x + 1$	**2)** $(11 - 4y)x$	**3)** $-2(x + 1)$
4) $(1 + y)x$	**5)** $3(x - 5)$	**6)** $3(1 + 3x)$
7) $8x - 3$	**8)** $2x(1 + 8y)$	**9)** $2(x - 20)$
10) $11x - 13$	**11)** $x + 4$	**12)** $2x(5y - 1)$
13) $3(x + 4)$	**14)** $2xy$	**15)** $3(x - 1)$
16) $0 = 2(2 - x)$	**17)** $3(x + 3)$	**18)** $(2 - y)5xy$

CHAPTER 3

Quick Exercises 1
1) $x = 15$
2) $y = 19$
3) $x = 16.2$
4) $n = 7.78$

Quick Exercises 2
1) $x = 4$
2) $n = 17$
3) $c = 44$
4) $y = 2.07$

Quick Exercises 3
1) $x = -\frac{1}{2}$
2) $x = \frac{17}{16}$
3) $x = 12$
4) $x = -\frac{1}{3}$

Quick Exercises 4
1) $y = 5\frac{2}{3}$
2) $x = -3$
3) $x = 15$
4) $y = \frac{1}{4}$

Quick Exercises 5
1) $n = -4$
2) $n = -16.4$
3) $y = 6$
4) $x = 5\frac{1}{2}$

Quick Exercises 6
1) $d = -1$
2) $c = -24.5$
3) $x = 18$
4) $x = -1.1$

Quick Exercises 7
1) $y = 2\frac{1}{2}$
2) $x = 1\frac{1}{3}$
3) $x = 22\frac{1}{2}$
4) $y = 12$

Quick Exercises 8
1) $y = \frac{1}{30}$
2) $x = -16$

Quick Exercises 9
1) $x = \frac{1}{3}$
2) $y = 5$
3) $y = -1$
4) $x = -2$

Quick Exercises 10
1) $m = 41$
2) $y = 15$
3) $x = 8$
4) $y = 44$

Quick Exercises 11
1) $m = 5$
2) $m = -4$
3) $m = 2$
4) $m = 3\frac{1}{2}$

Exercises 1
1) $-7\frac{2}{3}$
2) 5.7
3) 2.3
4) 1
5) -2.4
6) 3
7) $(\frac{1}{2})$
8) -4.3
9) 4.13
10) 4

Exercises 2
1) -21
2) 14
3) 1.95
4) 13
5) 15
6) 3
7) -3
8) 4

Exercises 3
1) 4.6
2) -7
3) 6.3
4) 5
5) 4
6) 1
7) 7.4
8) 14
9) 3.8
10) 4.8

ANSWERS

Exercises 4
1) 12 **2)** 4 **3)** 9.8 **4)** 24 **5)** 6 **6)** 20
7) 11.2 **8)** 5.8

Exercises 5
1) 1.7 **2)** 3.75 **3)** 1 **4)** $2\frac{1}{4}$ **5)** 4 **6)** -3
7) 1 **8)** -2 **9)** $13\frac{1}{3}$ **10)** $4\frac{2}{7}$

Exercises 6
1) -4 **2)** 3 **3)** 1.6 **4)** -9 **5)** 10 **6)** 10
7) 8 **8)** -4

Exercises 7
1) 2 **2)** 4 **3)** 2 **4)** 3 **5)** $231 **6)** 26.22
7) -48 **8)** 50

Exercises 8
1) -4 **2)** -5 **3)** -12 **4)** 9.6 **5)** 7 **6)** 20
7) 11.2 **8)** 8

Exercises 9
1) -18 **2)** 18 **3)** -1 **4)** -6 **5)** -7.2 **6)** 18
7) -31 **8)** 5 **9)** -1.8 **10)** -12

Exercises 10
1) -6 **2)** 3 **3)** 3 **4)** 6.8 **5)** -8 **6)** -2
7) 1.6 **8)** 3

Exercises 11
1) 8 **2)** -45 **3)** $3\frac{3}{11}$ **4)** $12\frac{4}{5}$ **5)** $-\frac{4}{11}$ **6)** $(\frac{3}{4})$
7) $-\frac{6}{35}$ **8)** 3

Exercises 12
1) 5 **2)** 9 **3)** 1 **4)** 12 **5)** 12 **6)** $\frac{1}{2}$
7) 5 **8)** -1

Exercises 13
1) $3261.54 - 453.17 = (3 \times 2)x$ **2)** $468 **3)** $150.30 + (50.45)(12) = x$ **4)** $755.70
5) $12 - 2 = (1/3)x + x$ **6)** 1 **7)** $(5/4)x + (5/2)x = 4$ **8)** $(9/4)$
9) $x + (x - 10) + (11 + x) = 76$ **10)** 25 **11** $49 = 2 + x + 23$ **12** 24

Exercises 14
1) -14 **2)** 9 **3)** -6 **4)** 7 **5)** -4 **6)** 42
7) 2.5 **8)** 2.5 **9)** -3 **10)** -8

Exercises 15
1) 4 **2)** -10 **3)** -3 **4)** -3 **5)** 3 **6)** -3
7) 8 **8)** 6 **9)** 9 **10)** -49

Exercises 16

1)	36.8	**2)**	15	**3)**	30	**4)**	6.3	**5)**	16	**6)**	1
7)	240	**8)**	22	**9)**	3	**10)**	8				

Exercises 17

1)	36	**2)**	-2.4	**3)**	0.25	**4)**	8	**5)**	15	**6)**	$-\frac{2}{7}$
7)	12	**8)**	1	**9)**	-17	**10)**	2				

Exercises 18

1)	4	**2)**	-5	**3)**	4	**4)**	12	**5)**	1	**6)**	4
7)	32	**8)**	2								

Exercises 19

1)	1	**2)**	$\frac{1}{2}$	**3)**	5	**4)**	-4	**5)**	5	**6)**	3
7)	10	**8)**	3	**9)**	4/3	**10)**	5/4				

Exercises 20

1)	9.6	**2)**	-1	**3)**	-30	**4)**	-6.3	**5)**	32	**6)**	-0.5
7)	6	**8)**	-2.0	**9)**	3	**10)**	2				

Exercises 21

1)	1	**2)**	12	**3)**	-6	**4)**	1	**5)**	$1\frac{2}{3}$	**6)**	-4
7)	-1	**8)**	0.5								

Exercises 22

1)	12	**2)**	6	**3)**	$\frac{2}{3}$	**4)**	$\frac{1}{16}$	**5)**	2	**6)**	$(\frac{1}{4})$
7)	1	**8)**	$(\frac{1}{40})$								

Exercises 23

1)	12	**2)**	7	**3)**	16	**4)**	12	**5)**	1	**6)**	4
7)	4	**8)**	1								

Exercises 25

1)	2	**2)**	$1\frac{1}{8}$	**3)**	$\frac{2}{3}$	**4)**	-6	**5)**	-5	**6)**	$(\frac{5}{9})$
7)	10	**8)**	$\frac{3}{4}$								

Exercises 26

1)	-7	**2)**	8	**3)**	-4	**4)**	4	**5)**	2	**6)**	33

Exercises 27

1)	$-2\frac{1}{2}$	**2)**	$-1\frac{1}{2}$	**3)**	$(\frac{1}{2})$	**4)**	$-\frac{7}{11}$	**5)**	$(\frac{5}{6})$	**6)**	$(\frac{5}{9})$
7)	-2	**8)**	$1\frac{6}{7}$	**9)**	$(\frac{3}{4})$	**10)**	4	**11)**	$2\frac{1}{8}$	**12)**	$1\frac{2}{3}$

Exercises 28

1)	36	**2)**	106	**3)**	176	**4)**	32	**5)**	47	**6)**	16
7)	20	**8)**	45	**9)**	20	**10)**	6				

Exercises 29

1)	11	**2)**	17	**3)**	1	**4)**	$3\frac{1}{2}$	**5)**	7	**6)**	6
7)	46	**8)**	4	**9)**	3	**10)**	8				

Exercises 30
1) 3   2) -15   3) $2\frac{1}{2}$ 4) -2   5) 18   6) 1/5

Exercises 31
1) 290   2) 90   3) 7,713.2   4) 0.6   5) 3.68   6) 5.8125
7) 42   8) 289.5   9) 19.2   10) 26

Exercises 32
1) $\left(\frac{21}{48}\right)$ 2) 1146   3) 2250   4) 35.2   5) 186   6) 728
7) 0.529   8) 153   9) 538   10) 81   11) 12.7   12) 6.58
13) 165.75   14) 800   15) 3   16) 7.6322   17) 3   18) 54

Exercises 33
1) 1.168   2) 11.68   3) 311.5   4) 10   5) 0.827   6) 0.99
7) 9   8) 12.96

Exercises 34
1) $\$5.00 + 0.50x = \$31.00$ 2) $29   3) 15.80

CHAPTER 4

Quick Exercises 1

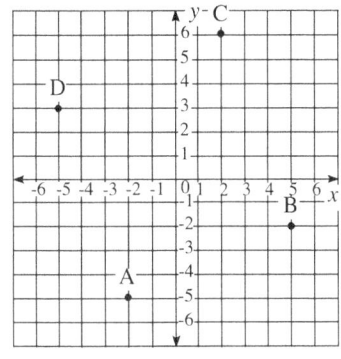

Quick Exercises 2
1) A(4, 4), B(4, 10)

2) 6 units

Quick Exercises 3
1) 14   2) -5   3) -1   4) -18

Quick Exercises 4
1)

x	−4	−1	2	5
y	−8	−2	4	10

2)

x	−2	0	2	3
y	−4	0	4	6

3)

x	−2	−1	0	1
y	2	1	0	−1

4)

x	−5	−1	3	11
y	5	1	−3	−11

Quick Exercises 5
1)

x	−4	−1	2	5
y	−11	−2	7	16

2)

x	−2	0	2	5
y	−5	1	7	16

3)

x	−4	−1	2	5
y	5	2	−1	−4

4)

x	0	−1	−4	−7
y	−5	1	7	13

Quick Exercises 6

1)

x	−2	0	2	4
y	−1	0	1	2

2)

x	−3	0	3	6
y	0	1	2	3

Quick Exercises 7
1) $y = 2x$ **2)** $y = 3x$ **3)** $y = -x$

Quick Exercises 8
1) $y = 3x + 2$ **2)** $y = 2x + 1$ **3)** $y = -x + 2$

Quick Exercises 9
1) $y = 3x - 2$ **2)** $y = 2x - 1$ **3)** $y = -2x - 1$

Quick Exercises 10
1) $y = \frac{1}{3}x$ **2)** $y = \frac{1}{3}x$ **3)** $y = -\frac{1}{2}x$

Quick Exercises 11
1) $y = \frac{1}{3}x + 2$ **2)** $y = \frac{1}{3}x - 1$ **3)** $y = -\frac{1}{2}(x - 1)$

Exercises 1
1) - 5)

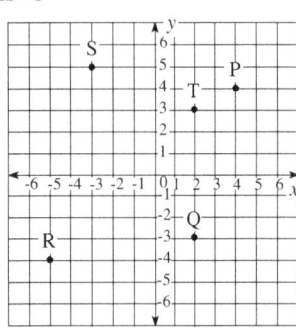

Exercises 2
1) A(2, 4) **2)** B(2, -2) **3)** C(0, 3) **4)** D(-6, 6) **5)** (5, 2) **6)** (-4, 4)

Exercises 3
1) B **2)** (3, 6) **3)** A **4)** (5, 3)

Exercises 4
1) (10, 7) **2)** (5, 30) **3)** (4, 10) (28, 4) **4)** $3, $7, $12

Exercises 5
1) C(3, 15) D(18, 15) **2)** 15 units **3)** B **4)** B to D

Exercises 6
1) 5 units **2)** 7 units **3)** 7 units **4)** 5 units

Exercises 7
1) 8 units **2)** 7 units **3)** 8 units **4)** 7 units

Exercises 8
1) 5 units **2)** 10 units **3)** 5 units **4)** 10 units

Exercises 9

	A	B	C	D	E
x	−3	−1	1	3	5
y	−5	−3	−1	1	3

Exercises 10
1) 0 **2)** 0 **3)** -9 **4)** 17 **5)** 8 **6)** 4

Exercises 11

1)

x	0	1	2	3
y	0	4	8	12

2)

x	−1	0	1	2
y	−3	0	3	6

3)

x	0	1	2	3
y	0	5	10	15

4)

x	−3	0	3	6
y	−3	0	3	6

5)

x	−1	0	1	2
y	−2	0	2	4

6)

x	−2	−1	0	1
y	−10	−5	0	5

7)

x	−3	0	3	6
y	9	0	−9	−18

8)

x	0	−2	−4	−6
y	0	8	16	24

Exercises 12

1)

x	0	1	2	3
y	−2	1	4	7

2)

x	−1	0	1	2
y	−1	1	3	5

3)

x	−1	0	1	2
y	−3	−1	1	3

4)

x	−3	0	3	6
y	−10	−1	8	17

Exercises 13

1)

x	0	1	2	3
y	−1	0	1	2

2)

x	0	1	2	3
y	1	0	−1	−2

3)

x	0	1	2	3
y	1	−3	−7	−11

4)

x	−1	1	3	5
y	−6	2	10	18

5)

x	−3	0	3	6
y	10	1	−8	−17

6)

x	0	1	2	3
y	−2	1	4	7

Exercises 14

1)

x	−6	−2	2	18
y	−3	−1	1	9

2)

x	−6	−3	0	15
y	−2	-1	0	5

3)

x	6	3	0	-15
y	-2	-1	0	5

4)

x	-4	0	4	20
y	-3	-1	1	9

Exercises 15

1)

x	6	4	2	-8
y	-2	-1	0	5

2)

x	4	0	-4	-20
y	-3	-1	1	9

3)

x	3	0	-3	-18
y	-2	-1	0	5

4)

x	-9	-6	-3	12
y	-2	-1	0	5

5)

x	-12	-9	-6	9
y	-2	-1	0	5

6)

x	12	9	6	-9
y	-2	-1	0	5

Exercises 16

1) $y = 5x$ **2)** $y = -2x$ **3)** $y = 6x$ **4)** $y = 2x$

Exercises 17

1) $y = 3x + 2$ **2)** $y = x + 4$ **3)** $y = 2x - 1$ **4)** $y = -x - 2$

Exercises 18

1) $y = -\frac{1}{3}x$ **2)** $y = -\frac{1}{4}x$ **3)** $y = \frac{1}{2}x$ **4)** $y = -\frac{1}{2}x$

Exercises 19

1) $y = -\frac{1}{3}x + 2$ **2)** $y = \frac{1}{4}x + 1$ **3)** $y = \frac{1}{2}x - 1$ **4)** $y = -\frac{1}{2}x - 2$

CHAPTER 5

Quick Exercises 1

$y = -3x$

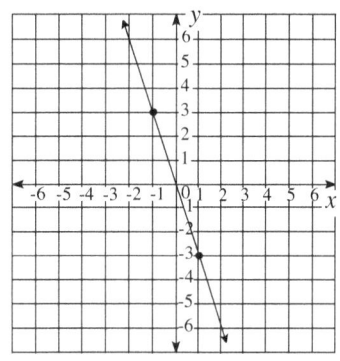

Quick Exercises 2

$y = x + 2$

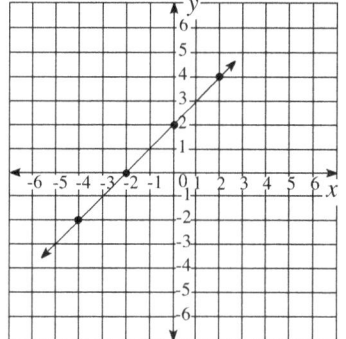

Quick Exercises 3

$y = -3x + 1$

x	-1	0	1	2
y	4	1	-2	-5

$y = x - 3$

x	-2	0	2	4
y	-5	-3	-1	1

Quick Exercises 4
1) Slope: -1, *x*-inter: 3, *y*-inter: 3 **2)** Slope: (1/2), *x*-inter: -2, *y*-inter: 1
3) Slope: 2, *x*-inter: 6, *y*-inter: -3 **4)** Slope: -(3/4), *x*-inter: 0, *y*-inter: 0

Quick Exercises 5
1) Slope: -(5/3),, *x*-inter: 3, *y*-inter: 5 **2)** Slope: (3/5), *x*-inter: 3, *y*-inter: -5

Quick Exercises 6
1) $y = 2x + 2$ **2)** $y = 3x - \frac{1}{2}$ **3)** $y = -\frac{1}{3}x + 1$ **4)** $y = -2x - 1$

Quick Exercises 7
1) $-\frac{10}{5}$ **2)** $-\frac{7}{11}$ **3)** 10 **4)** $\frac{5}{7}$

Quick Exercises 8
1) YES **2)** NO

Quick Exercises 9
1) YES **2)** NO

Quick Exercises 10
1) $y = \frac{1}{2}x + \frac{1}{2}$ **2)** $y = x + 2$ **3)** $y = -x + 1$ **4)** $y = -\frac{1}{2}x - \frac{1}{2}$
5) NO **6)** YES

Exercises 1
1) $y = 3x$ **2)** $y = 2x + 1$

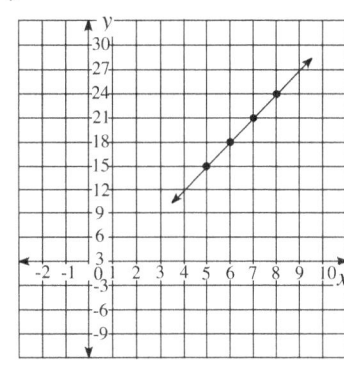 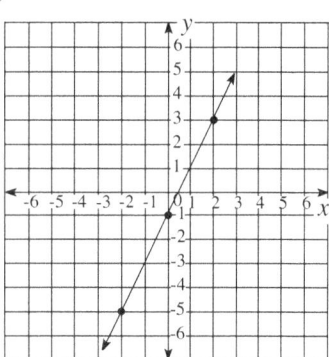

Exercises 2
1) $y = (1/3)x$ **2)** $y = (1/2)x$

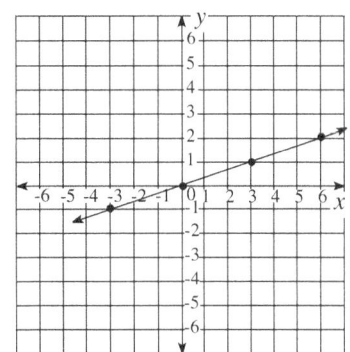 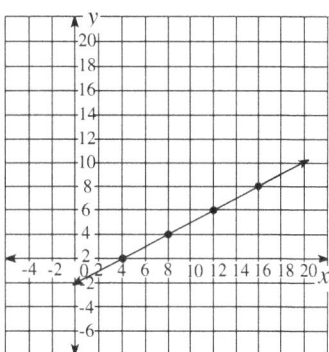

3) $y = 2x + 1$

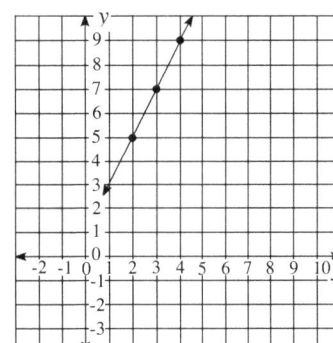

4) $y = 2x - 1$

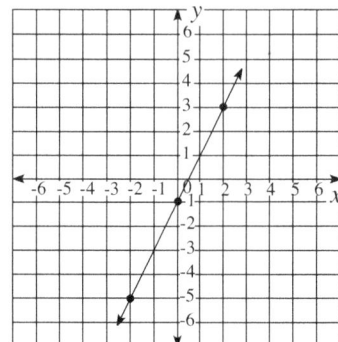

Exercises 3

1) $y = -2x - 1$

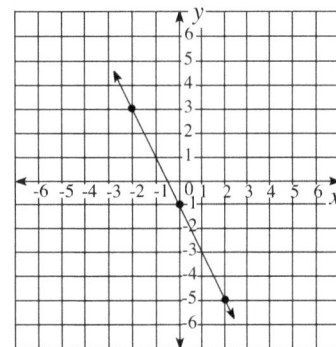

2) $y = 3x + 1$

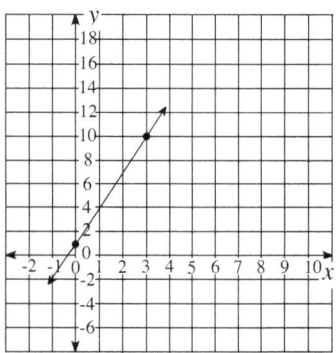

3) $y = (1/2)x - 1$

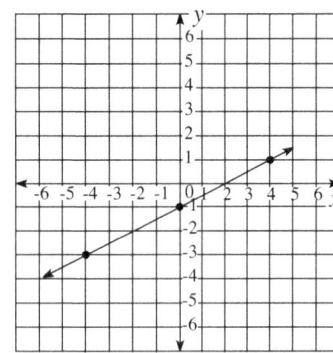

4) $y = -(1/3)x + 2$

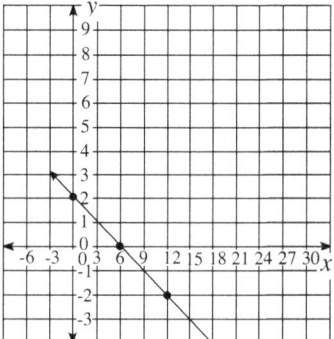

Exercises 4

1) $y = -3x$

x	−1	0	1	2
y	3	0	−3	−6

2) $y = 2x + 2$

x	−2	0	2	4
y	−2	2	6	10

3) $y = -(1/3)x - 2$

x	−6	−3	0	3
y	0	−1	−2	−3

4) $y = (2/3)x - 2$

x	−3	0	3	6
y	−4	−2	0	2

Exercises 5

1) x-inter: 1, y-inter: -1
2) x-inter: 1, y-inter: 2
3) x-inter: 1, y-inter: 4
4) x-inter: 1/2, y-inter: -3
5) x-inter: 1/3, y-inter: 2
6) x-inter: -5, y-inter: -(5/2)
7) x-inter: 0, y-inter: 0
8) x-inter: -3, y-inter: 3
9) x-inter: -1, y-inter: 0
10) x-inter: 8, y-inter: 2
11) x-inter: 15/2, y-inter: -3
12) x-inter: 0.1, y-inter: -0.5

Exercises 6

1) slope: 2, y-inter: 1
2) slope: -2, y-inter: -1
3) slope: 1, y-inter: 4
4) slope: -1, y-inter: 3
5) slope: -2, y-inter: -3
6) slope: 5, y-inter: 2
7) slope: 2, y-inter: 1
8) slope: 3, y-inter: -2
9) slope: 0, y-inter: -1
10) slope: -(1/2), y-inter: -3
11) slope: 1/9, y-inter: 2/3
12) slope: -(1/10), y-int: -0.5

Exercises 7

1) Slope: -1, x-inter: 2, y-inter: 2
2) Slope: -(4/5), x-inter: -5, y-inter: -4
3) Slope: -(5/3), x-inter: 1/2, y-inter: -2
4) Slope: -(5/2), x-inter: -2, y-inter: -5
5) Slope: (2/5), x-inter: 1/2, y-inter: 21/5
5) Slope: -(2/3), x-inter: 0, y-inter: 0

Exercises 8

1) $y = -\frac{1}{2}x + 1$
2) $y = -6x - 12$
3) $y = -\frac{2}{3}x + 4$
4) $y = \frac{1}{2}x - 8$
5) $y = \frac{5}{4}x + 10$
6) $y = \frac{1}{3}x + 4$

Exercises 9

1) $y = x + 1$
2) $y = 5x + 5$
3) $y = 2x + 2$
4) $y = x$
5) $y = x - 2$
6) $y = x - 1$
7) $y = 5$
8) *undefined*

Exercises 10

1) $y = 4x + 1$
2) $y = -x - 2$
3) $y = \frac{1}{3}x - \frac{2}{3}$
4) $y = -\frac{1}{2}x + 7$
5) $y = -3x + 4$
6) $y = x - 3$

Exercises 11

1) $-\frac{1}{2}$
2) $-\frac{5}{4}$
3) $+\frac{3}{8}$
4) $+\frac{1}{5}$
5) $3\frac{1}{3}$
6) $-1\frac{4}{5}$
7) $1\frac{1}{4}$
8) $-1\frac{3}{4}$

Exercises 12

1) -3
2) -3
3) 1
4) -3
5) $+\frac{3}{4}$
6) $\frac{1}{5}$
7) $1\frac{1}{4}$
8) $-1\frac{3}{4}$

Exercises 13

1) x-inter: -4, y-inter: 4
2) x-inter: -5, y-inter: 3(1/2)
3) x-inter: 3, y-inter: 3
4) x-inter: -4, y-inter: -4

Exercises 14

1) YES, SLOPE: 1
2) YES, SLOPE: -1
3) NO, S: -4, -3
4) YES, S: -(3/2)
5) YES, S: 1
6) YES, S: 2

Exercises 15

1) $y = 2$
2) $y = -3x - 19$
3) $y = -3x + 11$
4) $y = x + 4$
5) $y = x + 3$
6) $y = -\frac{1}{2}x - \frac{5}{8}$
7) $y = 3x + 3$
8) $y = -x - 6$

Exercises 16
1) $y = -3x + 7$
2) $y = \frac{1}{5}x + 7$
3) $y = \frac{5}{2}x + 7$
4) $y = -2x + 7$
5) $y = x + 7$
6) $y = x + 7$
7) $y = -\frac{3}{2}x + 7$
8) $y = \frac{1}{3}x + 7$

Exercises 17
1) $y = \frac{1}{4}x$
2) $y = -\frac{1}{3}x - \frac{19}{3}$
3) $y = x + 1$
4) $y = -\frac{1}{2}x - 3\frac{3}{4}$
5) $y = -\frac{1}{3}x + \frac{7}{3}$
6) $y = -\frac{1}{2}x - \frac{3}{8}$
7) $y = -\frac{1}{5}x + 1\frac{14}{15}$
8) $y = \frac{1}{3}x - 3\frac{1}{5}$

Exercises 18
1) YES
2) NO
3) NO
4) YES

Exercises 19
1) NO
2) YES
3) YES
4) NO
5) YES
6) NO
7) YES
8) NO

Exercises 20
1) YES
2) YES
3) YES
4) NO
5) NO
6) NO
7) NO
8) YES

Exercises 21
1) NO
2) NO
3) NO
4) YES
5) YES
6) YES
7) YES
8) YES

Self-Test, Page 8
1) C	2) B	3) A	4) D	5) C	6) D
7) D	8) B	9) B	10) D	11) C	12) C
13) C	14) D	15) C	16) B	17) B	18) C
19) C	20) B	21) B	22) A	23) C	24) A

Self-Test, Page 29
1) A	2) D	3) D	4) D	5) D	6) D
7) D	8) B	9) C	10) D	11) A	12) A
13) C	14) C	15) A	16) B	17) A	18) C
19) C	20) C	21) A	22) D	23) A	24) C
25) B	26) A	27) A	28) C	29) C	30) D
31) D	32) B	33) D	34) B	35) C	36) D

Self-Test, Page 37
1) C	2) A	3) A	4) C	5) C	6) C
7) C	8) B	9) A	10) C	11) B	12) B
13) D	14) D				

Self-Test, Page 49
1) B	2) C	3) A	4) C	5) D	6) B
7) B	8) D	9) C	10) D	11) D	12) B
13) A	14) C	15) B	16) B	17) A	18) D
19) C	20) C	21) B	22) B	23) D	24) B
25) C	26) B	27) D	28) B	29) A	30) A

Self-Test, Page 64
1) D	2) C	3) B	4) A	5) A	6) A

7) B	8) A	9) B	10) C	11) D	12) A
13) B	14) D	15) D	16) D	17) B	18) D
19) A	20) A	21) D	22) B	23) D	24) B
25) A	26) B	27) B	28) B	29) C	30) A
31) B	32) B	33) A	34) B	35) D	36) B
37) A	38) B	39) C	40) C	41) A	42) C
43) B	44) D	45) D	46) C	47) B	48) A

Self-Test, Page 77

1) A	2) C	3) C	4) B	5) C	6) B
7) C	8) D	9) B	10) B	11) A	12) B
13) D	14) B	15) C			

Self-Test, Page 83

1) A	2) A	3) D	4) D	5) C	6) B
7) D	8) A	9) D	10) C	11) B	12) B
13) D	14) C	15) C	16) D	17) B	18) A
19) D	20) D	21) C	22) C		

Self-Test, Page 92

1) A	2) B	3) B	4) C	5) C	6) A
7) A	8) A	9) C	10) C	11) A	12) D
13) C	14) B	15) C	16) D	17) C	18) D
19) C	20) C				

Self-Test, Page 103

1) C	2) C	3) C	4) D	5) C	6) D
7) A	8) C	9) B	10) D	11) D	12) C
13) B	14) C	15) B	16) C	17) B	

Self-Test, Page 114

1) B	2) C	3) A	4) A	5) D	6) A
7) D	8) C	9) A	10) D	11) B	12) C
13) C	14) D	15) B	16) C	17) B	18) C
19) B	20) B	21) C	22) D	23) D	24) C
25) C	26) A	27) B			

Self-Test, Page 127

1) B	2) D	3) B	4) A	5) D	6) A
7) D	8) B	9) B	10) B		

Self-Test, Page 141

1) C	2) C	3) C	4) A	5) C	6) C
7) C	8) A	9) C	10) B	11) D	12) B
13) D	14) C	15) A	16) D	17) A	18) A
19) B	20) D	21) A	22) B	23) D	

Self-Test, Page 153

1) D	2) B	3) C	4) B	5) D	6) A
7) D	8) B	9) D	10) B	11) D	12) C

Visit us at WWW.IQMATHS.com

ISBN: 978-1-5232673-6-1

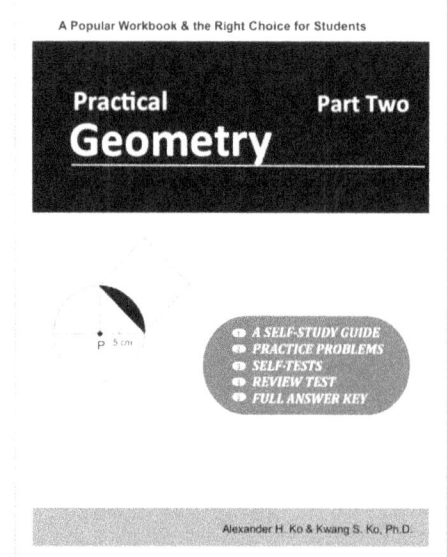

ISBN: 978-1-5233620-1-1

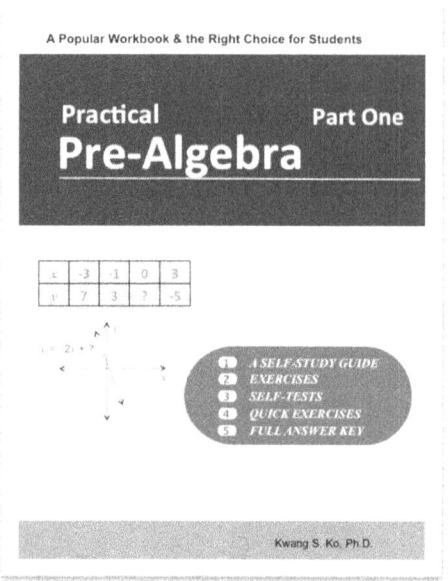

ISBN: 978-1-5233628-6-8

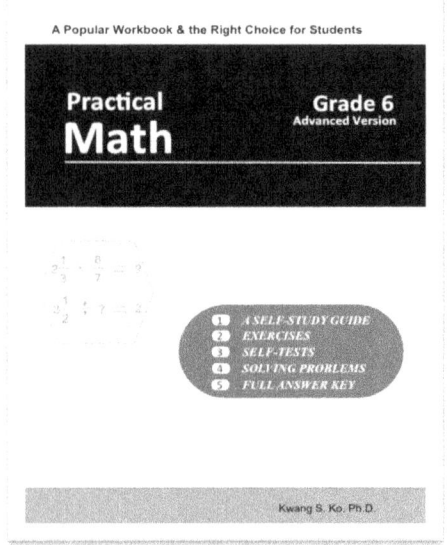

ISBN: 978-1-5233628-9-9

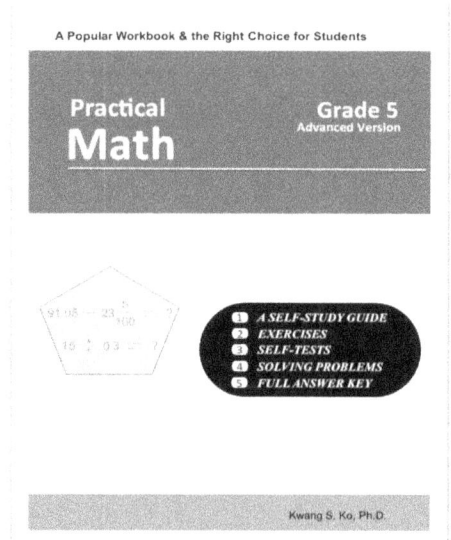

ISBN: 978-1-5233630-1-8

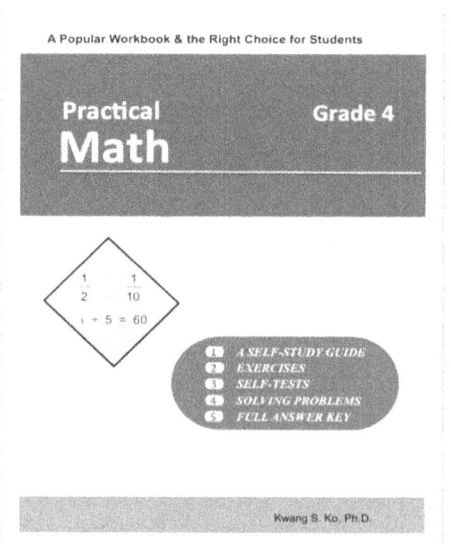

ISBN: 978-1-5233630-2-5

Other books are sold at WWW.IQMATHS.com.

www.ingramcontent.com/pod-product-compliance
Lightning Source LLC
Chambersburg PA
CBHW080011210526
45170CB00015B/1971